Institute of Mathematical Statistics

LECTURE NOTES-MONOGRAPH SERIES
Shanti S. Gupta, Series Editor

Volume 2

Survival Analysis

Proceedings of the Special Topics Meeting sponsored by the Institute of Mathematical Statistics, October 26–28, 1981, Columbus, Ohio

Edited by

John Crowley
Fred Hutchinson Cancer Research Center
University of Washington

Richard A. Johnson
University of Wisconsin, Madison

Institute of Mathematical Statistics
Hayward, California

Institute of Mathematical Statistics

Lecture Notes-Monograph Series

Series Editor, Shanti Gupta, Purdue University

Library of Congress Catalog Card Number: 82-84316

International Standard Book Number 0-940600-02-1

Printed in the United States of America

PREFACE

The 178th meeting of the Institute of Mathematical Statistics, held at Ohio State University in Columbus, Ohio, October 26-28, 1981, was organized as a Special Topics Meeting on Survival Analysis. The intent was to gather workers interested in the analysis of life length, from both reliability and biomedical applications, and to share progress and ideas across these disciplines. Survival analysis has been an active and exciting area of research for the past several decades, and one which is still gaining momentum today. The breadth of current activity in the field is illustrated in this Proceedings Volume, which includes invited papers from the meeting covering seven main topics:

I. <u>Counting Processes and Survival Analysis</u>

This first paper reviews the application of counting processes and the associated martingales to the large sample theory for a broad class of problems in survival analysis, from one and two sample situations to regression.

II. <u>Nonparametric Inference for a Single Sample</u>

Research on the one-sample problem is represented by three papers, covering a smooth version of the product-limit estimator, a generalization of the product-limit estimator to progressive censoring schemes, and an estimator of the hazard or failure rate.

III. <u>Proportional Hazards and Log-Linear Models</u>

Papers in this group give a comparison of estimators of the ratio of hazard functions in the context of a proportional hazards model, and a comparison of least squares and partial likelihood approaches when a log-linear model and proportional hazards both hold.

A new algorithm for least-squares type estimation for parameters in a linear model (possibly after transformation) for censored survival data is also given and investigated.

IV. Regression Approaches

Other research presented on regression includes an analysis of the statistical aspects of the inverse Gaussian model, and of the Box-Cox transformation toward normality with censored data. General considerations regarding the errors in variables problem are also discussed.

V. Problems in System Reliability

Inference procedures with time-truncated life-test data from an exponential model and a mixture of exponentials are given, and the properties of a system with imperfect repair of components derived. General limit theorems for a class of life-testing problems are also presented.

VI. Multivariate Distributions and Competing Risks

General notions regarding the concept of negative dependence of random variables are given, as well as some aspects of the theory of possibly dependent competing risks. Two papers present and explore estimators of the bivariate distribution function with censored observations, one using a model with exponential hazards, the other taking a nonparametric point of view.

VII. Group Sequential Methods in Clinical Trials

Both large sample and Monte Carlo approaches are used to investigate the properties of various statistics as applied at several times during the course of a clinical trial.

There were 25 invited papers as well as 6 sessions for contributed papers. The invited speakers were:

Per Kragh Andersen
Statistical Research Unit
Danish Medical and Social Research Councils

Richard Barlow
University of California, Berkeley

Asit P. Basu
University of Missouri, Columbia

Gouri K. Bhattacharyya
University of Wisconsin, Madison

Henry W. Block
University of Pittsburgh

Gregory Campbell
Laboratory of Statistical & Mathematical Methodology
National Institutes of Health

John Crowley
Fred Hutchinson Cancer Research Center
University of Washington

Kjell A. Doksum
University of California, Berkeley

T.R. Fleming
Mayo Clinic

Mitchell H. Gail
National Cancer Institute

Joseph C. Gardiner
Michigan State University

Richard A. Johnson
University of Wisconsin, Madison

J.D. Kalbfleisch
University of Walterloo

Jerome Klotz
Ohio State University

Sue Leurgans
University of Wisconsin, Madison

N.R. Mann
University of California, Los Angeles

Paul Meier
University of Chicago

Janet Myhre
Claremont McKenna College

Ross Prentice
Fred Hutchinson Cancer Research Center
University of Washington

Frank Proschan
Florida State University

Nozer D. Singpurwalla
George Washington University

V. Susarla
Michigan State University

Anastasios A. Tsiatis
Harvard University
Sidney Farber Cancer Institute

John Van Ryzin
Columbia University

Marvin Zelen
Harvard University
Sidney Farber Cancer Institute

In a few instances the papers included here differ somewhat from the remarks given at the meeting because of prior publication elsewhere.

ACKNOWLEDGMENTS

The Editors would first like to thank all who attended the Special Topics Meeting on Survival Analysis for making it a stimulating and rewarding experience. We gratefully acknowledge the support of the Institute of Mathematical Statistics for sponsoring the conference and for publishing this Proceedings Volume. Special thanks are due to Heebok Park and Bruce Trumbo for help with the production of the volume, and to John Klein, Mark Berliner, Doug Wolfe and Jerome Klotz for handling the arrangements at Columbus. Special credit is due the following people, who helped read and edit the manuscripts: Gouri Bhattacharyya, Shu-Mei Chen, Dennis Cox, Michael Jones, Sue Leurgans, Kung-Yee Liang, Barbara McKnight, Frances Olszewski, Barry Storer, Wei-Yann Tsai and Joseph G. Voelkel.

The excellent typing was done by Joy Hoggarth.

John Crowley
Fred Hutchinson Cancer Research Center
University of Washington

Richard A. Johnson
University of Wisconsin, Madison

TABLE OF CONTENTS

ON THE APPLICATION OF THE THEORY OF COUNTING PROCESSES IN THE STATISTICAL ANALYSIS OF CENSORED SURVIVAL DATA

Per Kragh Andersen

Statistical Research Unit, Danish Medical and Social Science Research Councils

0. SUMMARY

It was demonstrated by Aalen (1978) how the theory of multivariate counting processes gives a general framework in which both censored survival data and inhomogeneous Markov processes may be analyzed, and how by means of martingale central limit theory the asymptotic distribution for all the classical linear nonparametric two-sample tests and their generalizations to censored data may be derived. In this paper these results will be surveyed and further developed to both the case of the comparison of $k(\geq 2)$ distributions (see Andersen, Borgan, Gill & Keiding, 1981) and to the case of regression models for survival data (Cox, 1972; Andersen & Gill, 1981).

1. Introduction

In survival analysis one is interested in the distribution of the time T to some event, usually denoted death, and very often the object of a study is to relate this distribution to individual characteristics which in the simplest form are group indicators. Frequently statistical models for survival data are specified via the intensity or hazard function $\alpha(t)$ for T. The hazard function denotes the infinitesimal probability of dying at time t given survival up to time t, and hence $\alpha(t)$ may be interpreted as the rate at which the event in question occurs at time t.

A survival model is the simplest example of a Markov process model in that there are only two states, "alive" and "dead", with an intensity equal to $\alpha(t)$ of a transition from the former state to the latter. In more general Markov processes the basic parameters are the forces of transition between the states.

From these facts it seems obvious that the natural framework in which to analyze such phenomena is one where various types of events may happen during time and where the rate at which the events occur can be specified. One such framework can be introduced by the notation of a multivariate counting process.

A univariate point process is a countable random set of points on the real line, and a multivariate point process is a collection of, say k univariate processes. If $N_i(t)$ is defined as the number of points in $[0,1]$ from the ith process, then N_i can be thought of as counting the events of type i before t, and N_i is called the counting process corresponding to the ith point process. Let $(F_t)_{t\geq 0}$ be an increasing family of σ-algebras. One possibility would be to let F_t be the σ-algebra generated by the multivariate counting process $((N_1(s),\ldots,N_k(s)),\ s\in[0,t])$, but a larger family can also be considered. Note that the fact that (F_t) is increasing reflects that time moves in a certain direction. We shall assume that the limit

(1) $$\Lambda_i(t+) = \lim_{h\downarrow 0} \frac{1}{h} E(N_i(t+h) - N_i(t) | F_t)\ , \quad t\geq 0,\ i=1,\ldots,k\ ,$$

exists and we shall call the random process $\Lambda_i(t)$ the intensity process of N_i; this concept generalizes the notion of a hazard rate.

The idea of using counting process theory in the analysis of survival data and other Markov processes is due to Aalen (1975, 1978). There the so-called multiplicative intensity model was introduced, this statistical method being specified by assuming that the intensity process has the form

(2) $$\Lambda(t) = \alpha(t)\ Y(t), \quad t\geq 0\ .$$

Here $\alpha(t)$ is an unknown function and $Y(t)$ is an observable stochastic process adapted to $\mathcal{F}_{t-}$. In a survival study $\alpha(t)$ will be the hazard function and $Y(t)$ the number of individuals at risk just before time t, while in a more general Markov chain $\alpha(t)$ is a force of transition and $Y(t)$ is the number at risk just before time t for the transition in question.

The intensity property (1) of $\Lambda_i(t)$ is (up to regularity conditions) equivalent to the fact that the processes

$$(3) \qquad M_i(t) = N_i(t) - \int_0^t \Lambda_i(u)du \, , \quad t \geq 0 \, , \quad i=1,\ldots,k \quad ,$$

are martingales, i.e., $E(M_i(t)|\mathcal{F}_u) = M_i(u)$, $t > u$. This observation is the basis for making a unified approach to the proofs of asymptotic properties of many estimators and test statistics known from the survival data literature since these statistics can often be expressed as stochastic integrals with respect to martingales, and since furthermore central limit theorems and other properties of martingales are very well studied.

The typical feature of survival data is that one is not always able to observe all the lifetimes; rather, for some individuals it is only known that the true lifetime T_i exceeds some quantity t_i. We denote the corresponding observation a (right) censored observation. Another advantage of using the counting process description of survival data is that it accommodates fairly general censoring patterns (see e.g., Gill, 1980, Section 3.1, or Andersen, Borgan, Gill & Keiding, 1981, Sections 2D and 3D).

The rest of this paper contains examples from survival analysis of the use of the theory of counting processes, martingales and stochastic integrals. For a more detailed survey of the probabilistic background the reader is referred to Aalen (1978), Gill (1980) and Andersen, Borgan, Gill & Keiding (1981) and the references therein.

2. The One-Sample Situation

The simplest situation with censored data is the one-sample set-up, where out of n independent identically distributed lifetimes T_i, some are observed, but for the rest it is only known that they are larger than some times t_i. Let $X_i = \min(T_i, t_i)$ and $\delta_i = I(T_i \le t_i)$. From these data we want to estimate the distribution F of the T_i's. The product-limit estimator (Kaplan & Meier, 1958) of the survivorship functions $S = 1-F$ is given by

$$(4) \qquad \hat{S}(t) = \prod_{i:X_i \le t} \left(1 - \frac{\delta_i}{Y(X_i)}\right), t \ge 0 \quad ,$$

where $Y(t) = \#\{i:X_i \ge t\}$. Let $N(t) = \#\{i:X_i \le t, \delta_i = 1\}$. Then N is a counting process with intensity process $\alpha(t)\ Y(t)$ (cf. (2)), where $\alpha(t) = -\frac{d}{dt} \log S(t)$ is the hazard function (see Aalen (1978, Example 1)). It follows that $\hat{S}(t) = \prod_{u \le t} (1 - \frac{dN(u)}{Y(u)})$ and from this fact and from (3) it was noted by Aalen and Johansen (1978) (see also Gill, 1980, Lemma 3.2.1) that $\hat{S}(t)/\tilde{S}(t)$ is a martingale, where $\tilde{S}(t)$ converges to $S(t)$ at an exponential rate; hence the asymptotic properties when $n \to \infty$ of $\hat{S}$ also proved by Breslow and Crowley (1974) can be derived very simply (see Gill, 1980, 1981).

The Nelson estimator (Nelson, 1969, 1972) for the cumulative hazard function $\beta(t) = \int_0^t \alpha(u)du$ is given by

$$(5) \qquad \hat{\beta}(t) = \int_0^t \frac{dN(u)}{Y(u)} \ , \ t \ge 0 \quad ,$$

(Aalen, 1978, Section 6.1), and using (3) we see that $\hat{\beta}(t) - \beta(t)$ is a martingale (aside from a term that converges to zero at an exponential rate), being a stochastic integral of the process Y^{-1} with respect to a martingale; hence the asymptotic properties (see also Breslow & Crowley, 1974) can be found directly. Ramlau-Hansen (1981) used counting process and martingale techniques to study kernel function estimation

$$\hat{\alpha}(t) = (1/b) \int_0^\infty K((t-s)/b)\ d\hat{\beta}(s) \quad ,$$

of the hazard function $\alpha(t)$ itself rather than the integrated hazard $\beta(t)$. Here the kernel function K is non-negative with integral 1 and the window b is a positive parameter. Thus simple proofs of consistency and asymptotic normality of $\hat{\alpha}(t)$ were obtained.

We shall conclude this section by noting that the tests studied by Breslow (1975), Hyde (1977), Hollander and Proschan (1979) and Harrington and Fleming (1981) for comparing the distribution of the T_i's with a known distribution F_0 (with hazard function α_0, say) can be shown to have essentially the form

(6) $$Z(t) = \int_0^t L(u)\,(d\hat{\beta}(u) - \alpha_0(u)du)\ ,\quad t \geq 0\ ,$$

for various choices of the process L(u) (see Andersen, Borgan, Gill & Keiding, 1981, Section 4). It follows that under $H_0: F = F_0$, Z is a martingale, and from this fact the asymptotic distribution of the test statistics can be derived.

3. The k-Sample ($k \geq 2$) Situation

In this situation the problem is one of comparing the survival of k distinct groups. In each group i we have the Nelson estimator $\hat{\beta}_i(t)$ given by (5) for the cumulative hazard function $\beta_i(t) = \int_0^t \alpha_i(u)du$. Under the null hypothesis $H_0: \alpha_1 = \ldots = \alpha_k$ (=α, say) we can estimate $\beta(t) = \int_0^t \alpha(u)du$ by

(7) $$\hat{\beta}(t) = \int_0^t \frac{d\bar{N}(u)}{\bar{Y}(u)}\ ,\quad t \geq 0\ ,$$

where $\bar{N} = N_1 + \cdots + N_k$, $\bar{Y} = Y_1 + \cdots + Y_k$, $N_i(t)$ is the stochastic process counting the number of failures in group i in [0,1], and $Y_i(t)$ is the number at risk in group i at time t-. A general test statistic for H_0 based on the processes

(8) $$Z_i(t) = \int_0^t L(u)\,Y_i(u)\,(d\hat{\beta}_i(u) - d\hat{\beta}(u)),\ t \geq 0\ ,\quad i=1,\ldots,k\ ,$$

comparing the individual estimates $\hat{\beta}_i(t)$ with the common value $\hat{\beta}(t)$ was introduced by Andersen, Borgan, Gill & Keiding (1981, Section 3A). These authors proved that special choices of the stochastic process $L(t)$ correspond to various previously suggested test statistics from the literature. Thus, $L(t) = 1$ yields the logrank test (Peto & Peto; 1972), $L(t) = \overline{Y}(t)$ corresponds to the generalized Kruskal-Wallis test of Breslow (1970); $L(t) = (\overline{Y}(t))^{\rho}$ for ρ in $[0,1]$ corresponds to the family of statistics suggested by Tarone & Ware (1977); and $L(t) = [\hat{S}(t-)]^{\rho}$ (cf. (4)) gives the class of tests considered by Harrington & Fleming (1981) generalizing the Kruskal-Wallis type test of Prentice (1978), which is obtained for $\rho = 1$. The counting process formulation reveals that under H_0 we may write

$$(9) \qquad Z_i(t) = \int_0^t L(u)\left(dM_i(u) - \frac{Y_i(u)}{\overline{Y}(u)}\, d\overline{M}(u)\right), \quad t \geq 0, \quad i=1,\ldots,k ,$$

where $M_i(t) = N_i(t) - \int_0^t \alpha(u)\, Y_i(u)du$, and $\overline{M} = M_1 + \cdots + M_k$; hence Z_i is a martingale. This gives a general way of finding the asymptotic distribution of the test statistics. (See Crowley & Thomas (1975) for a derivation of the asymptotic distribution of the logrank test using a different approach).

In the case k=2, we can equivalently test for H_0 using the process

$$(10) \qquad Z(t) = \int_0^t K(u)\,(d\hat{\beta}_2(u) - d\hat{\beta}_1(u)), \quad t \geq 0 ;$$

see Aalen (1978, Section 7). As special cases of (10) we get the logrank test ($K = Y_1Y_2/(Y_1 + Y_2)$), the Wilcoxon test of Gehan (1965) ($K = Y_1Y_2$), and the test of Efron (1967) ($K = \hat{S}_1\hat{S}_2$). These tests were studied carefully by Gill (1980), who verified the conditions for normality for special censoring schemes and gave a discussion of efficiency properties of the tests.

The problem of estimating hazard ratios using counting process techniques in the two-sample model was discussed by Andersen (1981) following up the results of Crowley (1975). See also the paper by Crowley, Liu & Voelkel (this volume).

The use of the tests (8) and (10) in more general Markov processes was discussed by Aalen (1978), Aalen, Borgan, Keiding & Thormann (1980) and

Andersen & Rasmussen (1982). Examples of the applicability in analyses of Markov processes is found in Borgan (1980).

4. The Cox Regression Model

The semiparametric regression model of Cox (1972) specifies the hazard function of an individual i with (possibly time-dependent) covariates $z_i(t)$ to have the form

$$\alpha_i(t) = \lambda_0(t)\, e^{\beta_0' z_i(t)}, \quad t \geq 0 \quad . \tag{11}$$

Here β_0 is a vector of unknown regression coefficients and λ_0 is an unknown and unspecified underlying hazard function. The problem of estimating β_0 and λ_0 was discussed by Cox (1972,1975), Breslow (1972,1974), Kalbfleisch & Prentice (1973) and Tsiatis (1981a), and reviewed by Kalbfleisch & Prentice (1980).

The counting process formulation of (11) (cf. Andersen & Gill, 1981) specifies the intensity process Λ_i for the counting process N_i corresponding to the ith individual to have the form

$$\Lambda_i(t) = \lambda_0(t)\, e^{\beta_0' z_i(t)}\, Y_i(t), \quad t \geq 0 \quad , \tag{12}$$

where $Y_i(t) = 1$ if i is under observation at time t- and 0 otherwise. It was proven by Johansen (1981) that in an extended model, replacing the absolutely continuous measure $\Lambda_0(t) = \int_0^t \lambda_0(u)du$ by an arbitrary measure Λ on $[0,\infty)$ and allowing jumps with a Poisson-distributed size, the joint likelihood $L(\beta,\Lambda)$ for β_0 and $\Lambda_0(t)$ based on independent processes $N_1,\ldots,N_n$ is maximized for fixed β by

$$\hat{\Lambda}_0(t) = \int_0^t \frac{d\bar{N}(u)}{\sum_{j=1}^{n} Y_j(u)\, e^{\beta' z_j(u)}}, \quad t \geq 0 \quad . \tag{13}$$

The estimate (13) was also considered by Tsiatis (1981a), and by interpolating (13) between failure times we get the estimate of Breslow (1972, 1974). Inserting (13) in $L(\beta,\Lambda)$ yields a partially maximized likelihood $L_c(\beta) = \max_\Lambda L(\beta,\Lambda)$ proportional to the well known Cox's likelihood (cf. Cox 1972, 1975). Hence the Cox likelihood is an also reasonable basis for the estimation of β_0 in the counting process model (12). It was noted by Andersen & Gill (1981) that evaluated at the true value β_0 the score statistic $U(\beta) = \frac{d}{d\beta} \log L_c(\beta)$ has the form

$$U(\beta_0) = \sum_{i=1}^n \int_0^\infty z_i(u)\, dM_i(u) - \int_0^\infty \frac{\sum_{i=1}^n Y_i(u)\, z_i(u)\, e^{\beta_0' z_i(u)}}{\sum_{i=1}^n Y_i(u)\, e^{\beta_0' z_i(u)}}\, d\bar{M}(u) , \tag{14}$$

where M_i is the martingale $N_i(t) - \int_0^t \Lambda_i(u)du$ and $\bar{M} = M_1 + \cdots + M_n$. Hence, the process $U(\beta_0,t)$, obtained by replacing ∞ by t in (14), is a martingale, and this fact gives a simple way of proving asymptotic normality of the score statistic (see also Tsiatis, 1981b and Sen, 1981). In the usual way this result extends to a proof of asymptotic normality of the solution $\hat{\beta}$ to the likelihood equation $U(\beta) = 0$. (See Andersen & Gill, 1981, for details). Næs (1982) obtained the same results under stronger conditions using discrete time martingale results.

The weak convergence of $\hat{\Lambda}_0(\cdot) - \Lambda_0(\cdot)$ to a Gaussian process on a compact interval $[0,\tau]$ (Tsiatis, 1981a) can be obtained by first rewriting this difference as

$$\begin{aligned}\hat{\Lambda}_0(t) - \Lambda_0(t) = & \int_0^t \left(\frac{1}{\sum_{j=1}^n Y_j(u)\, e^{\hat{\beta}' z_j(u)}} - \frac{1}{\sum_{j=1}^n Y_j(u)\, e^{\beta_0' z_j(u)}} \right) d\bar{N}(u) \\ & + \left(\int_0^t \frac{d\bar{N}(u)}{\sum_{j=1}^n Y_j(u)\, e^{\beta_0' z_j(u)}} - \Lambda_0(t) \right) .\end{aligned} \tag{15}$$

The second term in (15) is (asymptotically equivalent to) a martingale, and by a Taylor expansion of the first term around β_0 this fact can be combined with the asymptotic normality of $\hat{\beta}$ to prove the weak convergence of $\hat{\Lambda}_0(\cdot) - \Lambda_0(\cdot)$.

An example of using the model (12) in a Markov process situation is found in Andersen & Rasmussen (1982) (see also Andersen & Gill, 1981).

Lustbader (1980) and Oakes (1981) showed how several well-known two-sample test statistics could be obtained as score tests from (11) by appropriate choices of time-dependent covariates. In fact, every k-sample test statistic of the form (8) can be obtained from (12) as a score test by letting $z_{ij}(t) = L(t)$ if individual i belongs to group j at time t and 0 otherwise.

ACKNOWLEDGEMENT

The author's participation in the IMS special topics meeting on survival analysis in Columbus, Ohio, 26-28 October 1981 was supported by the Danish Medical Research Council (Grant No. 12-2590) and the Danish Social Science Research Council (Grant No. 14-1997).

REFERENCES

Aalen, O.O. (1975). Statistical inference for a family of counting processes. Ph.D. dissertation. Department of Statistics, University of California, Berkeley.

Aalen, O.O. (1978). Nonparametric inference for a family of counting processes. Annals of Statistics **6**, 701-726.

Aalen, O.O., Borgan, Ø., Keiding, N. and Thormann, J. (1980). Interaction between life history events. Nonparametric analysis for prospective and retrospective data in the presence of censoring. Scandinavian Journal of Statistics **7**, 161-171.

Aalen, O.O. and Johansen, S. (1978). An empirical transition matrix for non-homogeneous Markov chains based on censored observations. Scandinavian Journal of Statistics 5, 141-150.

Andersen, P.K., Borgan, Ø., Gill, R.D. and Keiding, N. (1981). Linear non-parametric tests for comparison of counting processes, with applications to censored survival data. Research Report 81/4, Statistical Research Unit, Danish Medical and Social Science Research Councils. To appear in the International Statistical Review, 1982.

Andersen, P.K. and Gill, R.D. (1981). Cox's regression model for counting processes: a large sample study. Research Report 81/6, Statistical Research Unit, Danish Medical and Social Science Research Councils. To appear in the Annals of Statistics, 1982.

Andersen, P.K. and Rasmussen, N.K. (1982). Admissions to psychiatric hospitals among women giving birth and women having induced abortion. A statistical analysis of a counting process model. Research Report from Statistical Research Unit, Danish Medical and Social Science Research Councils, Copenhagen.

Andersen, P.K. (1981). Comparing survival distributions via hazard ratio estimates. Research Report 81/7, Statistical Research Unit, Danish Medical and Social Science Research Councils. To appear in the Scandinavian Journal of Statistics, 1983.

Borgan, Ø. (1980). Applications of non-homogeneous Markov chains to medical studies. Nonparametric analysis for prospective and retrospective data. In Explorative Datenanalyse. Frühjahrstagung, München, 1980. Proceedings (eds. N. Victor, W. Lehmacher and W. van Eimeren), pp. 102-115. Springer's series Medizinische Informatik und Statistik, Band 26.

Breslow, N.E. (1970). A generalized Kruskal-Wallis test for comparing K samples subject to unequal patterns of censorship. Biometrika 57, 579-594.

Breslow, N. (1972). Contribution to the discussion of the paper by D.R. Cox. Journal of the Royal Statistical Society B 34, 187-220.

Breslow, N. (1974). Covariance analysis of censored survival data. Biometrics 30, 89-99.

Breslow, N.E. (1975). Analysis of survival data under the proportional hazards model. International Statistical Review 43, 45-58.

Breslow, N.E. and Crowley, J. (1974). A large sample study of the life table and product limit estimates under random censorship. Annals of Statistics 2, 437-453.

Cox, D.R. (1972). Regression models and life tables (with discussion). Journal of the Royal Statistical Society B 34, 187-220.

Cox, D.R. (1975). Partial likelihood. Biometrika 62, 269-276.

Crowley, J. (1975). Estimation of relative risk in survival studies. Technical Report No. 423. Department of Statistics, University of Wisconsin-Madison.

Crowley, J. and Thomas, D.R. (1975). Large sample theory for the logrank test. Technical Report No. 415. Department of Statistics, University of Wisconsin-Madison.

Efron, B. (1967). The two sample problem with censored data. Proceedings of the Fifth Berkeley Symposium on Mathematical Statistics and Probability, Vol, IV, University of California Press, Berkeley, California, 831-853.

Gehan, E.A. (1965). A generalized Wilcoxon test for comparing arbitrarily singly-censored samples. Biometrika 52, 203-223.

Gill, R.D. (1980). Censoring and stochastic integrals. Mathematical Centre Tracts 124, Mathematisch Centrum, Amsterdam.

Gill, R.D. (1981). Large sample behaviour of the product-limit estimator on the whole line. Preprint 74/81, Mathematical Centre, Amsterdam. To appear in the Annals of Statistics, 1983.

Harrington, D.P. and Fleming, T.R. (1981). A class of rank test procedures for censored survival data. Technical Report Series No. 12, Section of Medical Research Statistics, Mayo Clinic. To appear in Biometrika, 1983.

Hollander, M. and Proschan, F. (1979). Testing to determine the underlying distribution using randomly censored data. Biometrics 35, 393-401.

Hyde, J. (1977). Testing survival under right censoring and left truncation. Biometrika 64, 225-230.

Johansen, S. (1981). An extension of Cox's regression model. Preprint No. 11, Institute of Mathematical Statistics, University of Copenhagen. To appear in the International Statistical Review, 1983.

Kalbfleisch, J.D. and Prentice, R.L. (1973). Marginal likelihoods based on Cox's regression and life model. Biometrika 60, 267-278.

Kalbfleisch, J.D. and Prentice, R.L. (1980). The Statistical Analysis of Failure Time Data. Wiley, New York.

Kaplan, E.L. and Meier, P. (1958). Nonparametric estimation from incomplete observations. Journal of the American Statistical Association 53, 457-481.

Lustbader, E.D. (1980). Time-dependent covariates in survival analysis. Biometrika 67, 697-698.

Nelson, W. (1969). Hazard plotting for incomplete failure data. Journal of Quality Technology 1, 27-52.

Nelson, W. (1972). Theory and application of hazard plotting for censored failure data. Technometrics 14, 945-966.

Næs, T. (1982). The asymptotic distribution of the estimator for the regression parameter in Cox's regression model. Scandinavian Journal of Statistics 9, 107-116.

Oakes, D. (1981). Survival times: aspects of partial likelihood (with discussion). International Statistical Review 49, 235-264.

Peto, R. and Peto, J. (1972). Asymptotically efficient rank invariant test procedures (with discussion). Journal of the Royal Statistical Society A 135, 185-206.

Prentice, R.L. (1978). Linear rank tests with right censored data. Biometrika 65, 167-179.

Ramlau-Hansen, H. (1981). Smoothing counting process intensities by means of kernel functions. Working Paper No. 43, Laboratory of Actuarial Mathematics, University of Copenhagen. To appear in the Annals of Statistics, 1983.

Sen, P.K. (1981). The Cox regression model, invariance principles for some induced quantive processes and some repeated significance tests. Annals of Statistics 9, 109-121.

Tarone, R.E. and Ware, J. (1977). On distribution-free tests for equality of survival distributions. Biometrika 64, 156-160.

Tsiatis, A.A. (1981a). A large sample study of Cox's regression model. Annals of Statistics 9, 93-108.

Tsiatis, A.A. (1981b). The asymptotic distribution of the efficient scores test for the proportional hazards model caclulated over time. Biometrika 68, 311-315.

SPLINE SMOOTH ESTIMATES OF SURVIVAL

Jerome Klotz

Ohio State University

1. Introduction

Let X be survival time with continuous distribution $F_X(x)$ and density $f_F(x)$. Similarly, let Y be time to censoring, independent of X, with continuous distribution $F_Y(y)$ and density $f_Y(y)$. We observe time on trial, T , and death or censoring indicator, D , where

$$T = \min(X,Y)$$
$$D = \begin{cases} 1 & \text{if } X \leq Y \quad \text{(death)} \\ 0 & \text{if } X > Y \quad \text{(censoring)} \end{cases} .$$

Using a sample $\{T_i, D_i); i=1,2,\ldots,n\}$ we wish to find a smooth estimate of the survival distribtion $1 - F_X(x) = P[X > x]$.

Define the hazard function by

$$h_X(x) = f_X(x)/(1 - F_X(x))$$

and the integrated hazard function by

$$H_X(x) = \int_0^x h_X(u)du = - \int_0^x d\,\ell n\,(1 - F_X(u)) \quad ,$$

which is related to survival by $1 - F_X(x) = e^{-H_X(x)}$. Defining the indicator function I[A] (1 or 0 according as the event A holds or not), the sample cumulative distribution is

$$F_n(t) = \sum_{i=1}^{n} I[T_i \le t] / n .$$

We are concerned with a smooth approximation of the hazard function over subintervals using polynomials. We write the polynomials as linear combinations of B-splines defined on the knots or points defining the subintervals. The B-spline or order r or polynomial of degree r-1 is defined for the non-decreasing sequence of knots

(1) $$\tau_{-r+1}, \tau_{-r+2}, \ldots, \tau_0, \tau_1, \ldots, \tau_K, \tau_{K+1}, \ldots, \tau_{K+r-1} ,$$

using the following divided differences:

$$g_r(\tau_j;t) = (\tau_j - t)_+^{r-1} = [\max(0, \tau_j - t)]^{r-1}$$

$$g_r(\tau_j, \tau_{j+1};t) = [g_r(\tau_{j+1};t) - g_r(\tau_j;t)] / (\tau_{j+1} - \tau_j)$$

$$\vdots$$

$$g_r(\tau_j, \tau_{j+1}, \ldots, \tau_{j+r}, t) = \left[\frac{g_r(\tau_{j+1}, \ldots, \tau_{j+r};t) - g_r(\tau_j, \ldots, \tau_{j+r-1};t)}{(\tau_{j+r} - \tau_j)}\right] .$$

Then the normalized B-spline is

$$N_{jr}(t) = (\tau_{j+r} - \tau_j)\, g_r(\tau_j, \tau_{j+1}, \ldots, \tau_{j+r};t) .$$

In case some knots coincide, continuity can be used for the definition. For a discussion of B-splines see de Boor (1976). Figure 1 gives graphs for $r = 2,3$ corresponding to linear and quadratic B-splines.

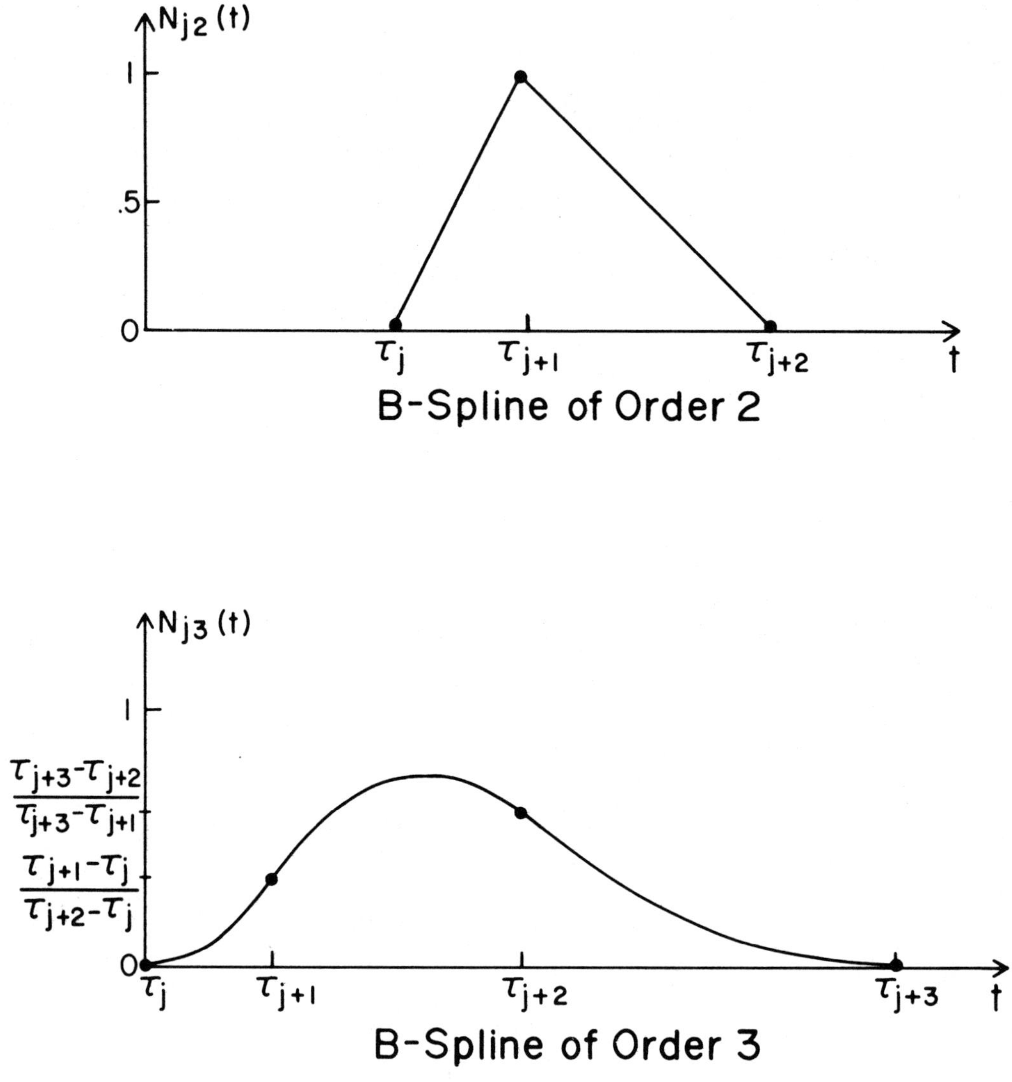

FIGURE 1. Linear and quadratic B-splines for knots $\{t_j\}$.

2. Hazard Approximation by Splines with Fixed Knots

We fit the model

$$h_X(x) = \sum_{j=-r+1}^{K-1} \theta_j N_{j,r}(x) \qquad (2)$$

over the interval $0 \le x \le \tau_K$ by selecting knots

$$\tau_{-r+1} = \tau_{-r+2} = \cdots = \tau_0 = 0 \le \tau_1 \le \tau_2 \le \cdots \le \tau_K = \tau_{K+1} = \cdots = \tau_{K+r-1} .$$

Although the model is parametric with parameters $\underset{\sim}{\theta} = (\theta_{-r+1}, \theta_{-r+2}, \ldots, \theta_{K-1})$, there is great flexibility through the choice of knots $\{\tau_k\}$ and spline order r.

We consider estimating $\underset{\sim}{\theta}$ by maximizing the likelihood. The joint continuous-discrete density under the random censorship model is

$$f_{T,D}(t,d) = [f_X(t)(1-F_y(t))]^d \, [f_Y(t)(1-F_X(t))]^{1-d}$$

$$= (h_X(t))^d \, (h_Y(t))^{1-d} \, (1-F_T(t)) \quad ,$$

where $1-F_T(t) = (1-F_X(t))\,(1-F_Y(t))$ by independence. The log-likelihood is then

$$\sum_{i=1}^{n} \ln f_{T,D}(t_i,d_i) = \sum_{i=1}^{n} [d_i \ln h_X(t_i) + \ln(1-F_X(t_i))]$$

$$+ \sum_{i=1}^{n} [(1-d_i) \ln h_Y(t_i) + \ln(1-F_Y(t_i))] \quad . \qquad (3)$$

Differentiating (3) with respect to $\underset{\sim}{\theta}$ using (2) gives

$$\frac{\partial}{\partial \underset{\sim}{\theta}} \sum_{i=1}^{n} \ln f_{T,D}(t_i,d_i) = \sum_{i=1}^{n} [d_i \underset{\sim}{N}_r(t_i) / (\underset{\sim}{N}_r(t_i) \underset{\sim}{\theta}') - \int_0^{t_i} \underset{\sim}{N}_r(u)du] \quad ,$$

where $\underset{\sim}{N}_r(x) = (N_{-r+1,r}(x), N_{-r+2,r}(x), \ldots, N_{K-1,r}(x))$ and $\underset{\sim}{\theta}'$ is the transpose of $\underset{\sim}{\theta}$.

If the solution of the derivative equation, $\partial \sum_i \ln f_{T,D}(t_i, d_i)/\partial \underset{\sim}{\theta} = \underset{\sim}{0}$, gives a nonnegative function $\hat{h}_X(x) = \sum_{j=-r+1}^{K-1} \hat{\theta}_j N_{j,r}(x)$ then we propose the estimator $1 - \hat{F}_X(x) = \exp(-\hat{H}_X(x))$, where $\hat{H}_X(x) = \int_0^x \hat{h}_X(u)du$.

Because of the necessity of choosing the degree r as well as suitable knots $\{\tau_k\}$ and then solving a messy non-linear derivative equation which we can only hope has a non-negative solution $\tilde{h}_X$, we turn instead to a simplification.

3. An Ad Hoc Estimator

The model $h_X(x) = \underset{\sim}{N}_r(x)\, \underset{\sim}{\theta}'$ breaks down when the knots defining $\underset{\sim}{N}_r$ are random variables. Nevertheless, motivated by the success of the estimator of Breslow (1974) that uses constant splines over random death times, we propose a similar simplication using linear splines $(r = 2)$. Specifically, we replace the knots $0 \le \tau_1 \le \tau_2 \le \cdots \le \tau_K$ in (1) by distinct death times $0 < T_{<1>} < T_{<2>} < \cdots < T_{<K>}$ which are different sorted values of T_i for which $D_i = 1$, $i = 1, 2, \ldots, n$. Using $N_{j,2}(T_{<j+1>}) = 1$, and 0 at other knots gives the minimizing solution $\tilde{\theta}_{-1} = 0$ and

$$\tilde{\theta}_j = m_{j+1} / \sum_{i=1}^{n} \int_0^{t_i} N_{j,2}(u)du \ , \quad \text{for } j = 0,1,2,\ldots,K-1 \ ,$$

where m_k is the number of death times equal $T_{<k>}$. Then the estimate of the integrated hazard is

$$\tilde{H}_X(x) = \sum_{k=1}^{K} \{m_k \int_0^x N_{k-1,2}(u)du \Big/ \sum_{i=1}^{n} \int_0^{t_i} N_{k-1,2}(t)dt\} \ .$$

From the identity

$$\int_0^x N_{k-1,2}(u)du = \left(\sum_{j \ge k-1} N_{j,3}(x))\ (T_{<k+1>} - T_{<k>} \right) / 2 \ ,$$

we can cancel the nonzero factor $(T_{<k+1>} - T_{<k>})/2$ from both numerator and denominator to obtain

$$(4) \qquad \tilde{H}_X(x) = \sum_{k=1}^{K} \{m_k \sum_{j \geq k-1} N_{j,3}(x) \Big/ \sum_{i=1}^{n} \sum_{r \geq k-1} N_{r,3}(t_i)\} \quad .$$

For computing, with knots $\tau_{k-1} < \tau_k < \tau_{k+1}$, we have

$$\sum_{j \geq k-1} N_{j,3}(x) = \begin{cases} 0, \text{ for } x \leq \tau_{k-1} \\ (x - \tau_{k-1})^2/((\tau_k - \tau_{k-1})(\tau_{k+1} - \tau_{k-1})), \text{ for } \tau_{k-1} \leq x < \tau_k \\ 1 - (\tau_{k+1} - x)^2/((\tau_{k+1} - \tau_k)(\tau_{k+1} - \tau_{k-1})), \text{for } \tau_k \leq x \leq \tau_{k+1} \\ 1, \text{ for } x \geq \tau_{k+1} \quad . \end{cases}$$

The estimator (4) is a non-negative differentiable monotone increasing function of x on the interval $[0, T_{<k>}]$ and thus

$$1 - \tilde{F}_X(x) = e^{-\tilde{H}_X(x)}$$

is a differentiable monotone decreasing function on this interval. Figure 2 gives an example of the estimator contrasted with the Kaplan-Meier estimator (1958).

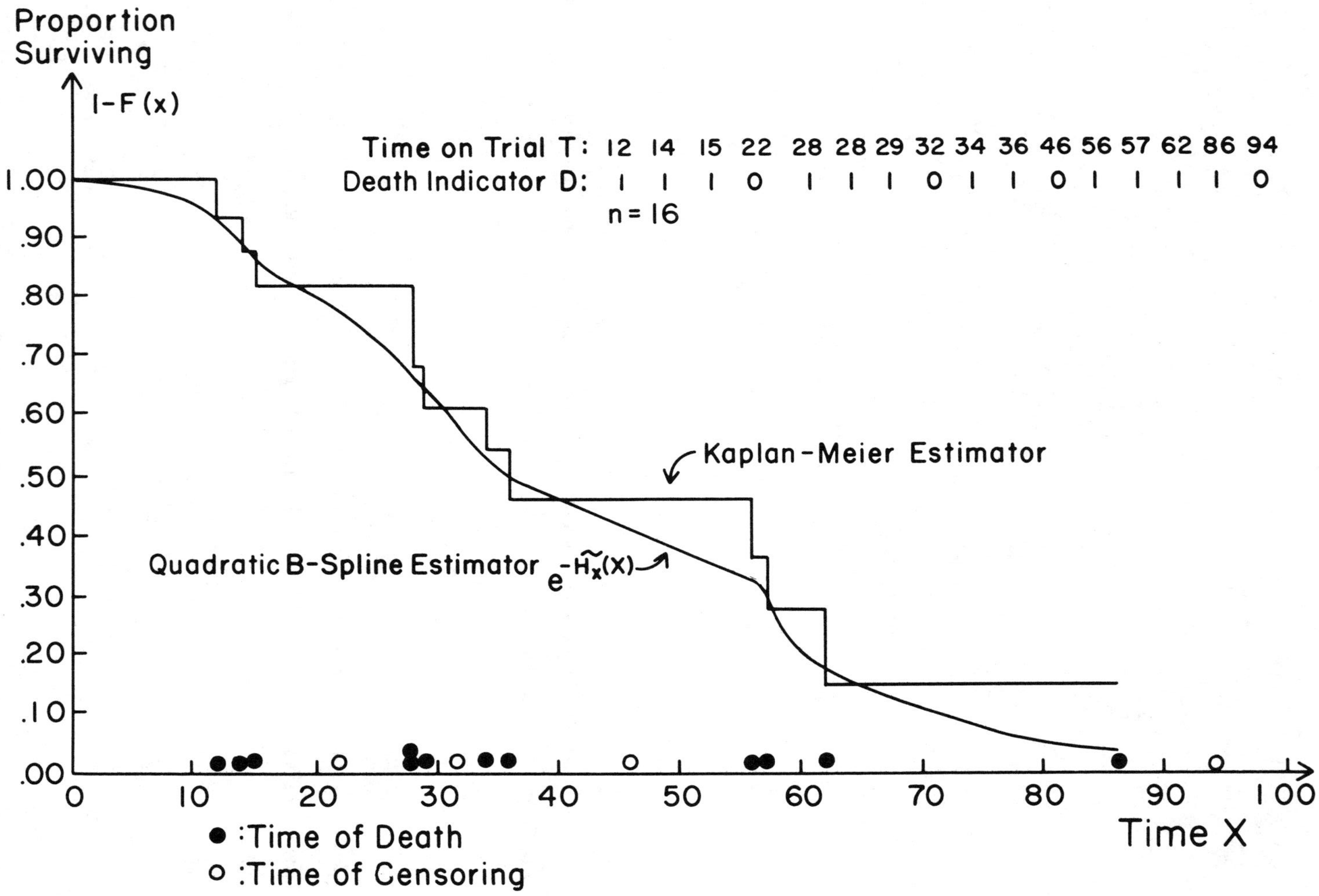

FIGURE 2. Comparison of the Kaplan-Meier estimator and the quadratic B-spline estimator.

4. Consistency of $\tilde{H}_X(x)$

The following theorem gives consistency of $\tilde{H}_X(x)$ and consequently $1-\tilde{F}_X(x)$ under some assumptions.

THEOREM 1:

If $f_X(t) > 0$ a.e. on the interval of t values for which $F_T(t) < 1$, $\tilde{H}_X(x) \overset{P}{\to} H_X(x)$ as $n \to \infty$ for x in the interior of the interval.

PROOF:

From equation (5) we obtain the inequalities

$$(6) \qquad I[\tau_{k+1} \leq x] \leq \sum_{j \geq k-1} N_{j,3}(x) \leq I[\tau_{k-1} < x] \quad .$$

By the continuity of F_X and F_Y, F_T is continuous and the ordered times on trial $0 < T_{(1)} < T_{(2)} < \cdots < T_{(n)}$ are distinct with probability 1. Consequently the ordered death times $0 < T_{[1]} < T_{[2]} < \cdots < T_{[M]}$ where $M = \sum_{i=1}^{n} D_i$ are distinct with probability one and $K = M$. Thus, $T_{<k>} = T_{[k]}$ and we have

$$L_n(x) \leq \tilde{H}_X(x) \leq U_n(x) \quad ,$$

where the upper bound

$$(7) \qquad U_n(x) = \sum_{k=1}^{M} I[T_{[k-1]} < x] \,/\, (n(1 - F_n(T^{-}_{[k+1]})))$$

is obtained by replacing the numerator and denominator of (4) by upper and lower bounds in (6) with knots $\{T_{[k]}\}$. Similarly

$$(8) \qquad L_n(x) = \sum_{k=1}^{M} I[T_{[k+1]} \leq x] \,/\, (n(1 - F_n(T_{[k-1]})))$$

is obtained from lower and upper bounds in the numerator and denominator of (4). Here we use the conventions $T_{[0]} = 0$ and $T_{[M+1]} = T_{[M]}$ in (7) and (8) so that (6) holds for $k = 1$ and $k = M$. Intuitively, the bounds $U_n(x)$ and $L_n(x)$ will be close to the empirical integrated hazard function

$$H_n(x) = \sum_{k=1}^{M} I[T_{[k]} < x] \,/\, (n(1 - F_n(T_{[\bar{k}]})))$$

$$= \sum_{i=1}^{n} \{D_i \, I[T_i < x] \,/\, (n(1 - F_n(T_i -)))\}$$

shown by Breslow and Crowley (1974) to converge weakly to

$$H_x(x) = E\{D \, I[T < x] \,/\, (1 - F_T(t))\}$$

using methods of Billingsly (1968). Consistency will follow by showing $U_n(x) - H_n(x) \overset{P}{\to} 0$ and $H_n(x) - L_n(x) \overset{P}{\to} 0$. We show convergence for the upper bound; the argument for the lower bound is similar. Write

$$U_n(x) - H_n(x) = \frac{1}{n} \sum_{k=1}^{M} \left\{ \frac{I[T_{[k-1]} < x]}{1 - F_n(T_{[\overline{k+1}]})} - \frac{I[T_{[k]} < x]}{1 - F_n(T_{[\bar{k}]})} \right\}$$

(9) $$= \sum_{k=1}^{M} \frac{I[T_{[k-1]} < x \le T_{[k]}]}{n(1 - F_n(T_{[\overline{k-1}]}) - w_{nk} - w_{nk+1})} +$$

$$\sum_{k=1}^{M} \frac{I[T_{[k]} < x] \, w_{nk+1}}{n(1 - F_n(T_{[\bar{k}]}))(1 - F_n(T_{[\bar{k}]} - w_{nk+1})} ,$$

where $w_{nk} = F_n(T_{[\bar{k}]}) - F_n(T_{[\overline{k-1}]})$. The expression (9) is in turn bounded by

$$[n(1 - F_n(x) - 2w_n^*)]^{-1} + H_n(x) \, w_n^* \,/\, (1 - F_n(x) - w_n^*) ,$$

where $w_n^* = \max_{1 \le k \le M} w_{nk}$. Since $\bar{F}_n(x) \overset{P}{\to} F_T(x) < 1$ and $H_n(x) \overset{P}{\to} H_X(x) < \infty$ we complete the proof by showing $w_n^* \overset{P}{\to} 0$. We bound w_{nk} by

$$w_{nk} = F_n(T_{[k]}) - F_n(T_{[k-1]})$$

$$= F_T(T_{[k]}) - F_T(T_{[k-1]}) +$$

$$(F_n(T_{[k]}) - F_T(T_{[k]})) - (F_n(T_{[k-1]}) - F_T(T_{[k-1]}))$$

$$\le F_T(T_{[k]}) - F_T(T_{[k-1]}) + 2 \sup(F_n(t) - F_T(t)) \quad .$$

Using the Glevenko Cantelli Theorem, $\sup(F_n(t) - F_T(t)) \overset{P}{\to} 0$, and so we show $\max_{1 \le k \le M} F_T(T_{[k]}) - F_T(T_{[k-1]}) \overset{P}{\to} 0$. Now for ε, $\delta > 0$,

$$P[\max_{1 \le k \le M} (F_T(T_{[k]}) - F_T(T_{[k-1]})) > \varepsilon]$$

$$\le \sum_{n(p-\delta) \le m \le n(p+\delta)} P[\max_{1 \le k \le M} (F_T(T_{[k]}) - F_T(T_{[k-1]})) > \varepsilon | M = m]\ P[M=m]$$

(10) $$+ P[|M - np| > n\delta]$$

$$\le \max_{n(p-\delta) \le m \le n(p+\delta)} P[\max_{1 \le k \le m} (F_T(T^*_{(k)}) - F_T(T^*_{(k-1)})) > \varepsilon]$$

$$+ P[|M - np| > n\delta] \quad ,$$

where M has a binomial (n,p) distribution, $p = P[X \le Y]$, and $T^*_{(1)}, \ldots, T^*_{(m)}$ are order statistics for an independent sample of size m from the distribution

$$F_*(t) = F_{T|D}(t|1)$$

with density

(11) $$f_*(t) = f_{T|D}(t|1) = f_X(t)\,(1 - F_Y(t))/p \quad .$$

The 2nd term in (10) goes to zero as $n \to \infty$ and so it remains to show $\max_{1 \le k \le m} (F_T(T^*_{(k)}) - F_T(T^*_{(k-1)})) \overset{P}{\to} 0$, as $m \to \infty$.

By the assumptions, we see from (11) that $F_*(t)$ is continuous in addition to $F_T(t)$ and we can write

$$\max_{1 \le k \le m} [F_T(T^*_{(k)}) - F_T(T^*_{(k-1)})]$$

$$= \max_{1 \le k \le m} [F_T(F_*^{-1}(F_*(T^*_{(k)}))) - F_T(F_*^{-1}(F_*(T^*_{(k-1)})))]$$

$$= \max_{1 \le k \le m} [F_T(F_*^{-1}(U_{(k)})) - F_T\, F_*^{-1}(U_{(k-1)})] \;,$$

where $U_{(k)}$ are order statistics from a uniform (0,1) distribution. Thus, by continuity it remains to show

$$\max_{1 \le k \le m} (U_{(k)} - U_{(k-1)}) \overset{P}{\to} 0 \quad \text{as } m \to \infty.$$

Finally, by properties of uniform order statistics,

$$P[\max_{1 \le k \le m} (U_{(k)} - U_{(k-1)}) > \varepsilon] \le \sum_{k=1}^{m} P[U_{(k)} - U_{(k-1)} > \varepsilon]$$

$$= m\, P[U_{(1)} > \varepsilon] = m(1-\varepsilon)^m \to 0 \quad \text{as } m \to \infty \quad .$$

It is likely that these strong assumptions can be weakened for proving consistency. However, some control on the spacings of adjacent death times may be required around the point x as $\tilde{H}_X(x)$ is a function of the order statistics and cannot be written as a counting process.

ACKNOWLEDGEMENT

Research was supported in part by the National Institutes of Health Grant No. 2-R01-CA-18332-07.

REFERENCES

Billingsly, P. (1968). Weak Convergence of Probability Measures. J. Wiley, New York.

Breslow, N. (1974). Covariance analysis of censored survival data. Biometrics **30**, 89-99.

Breslow, N. and Crowley, J. (1974). A large sample study of the life tables and product limit estimates under random censorship. Annals of Statistics **3**, 437-453.

de Boor, C. (1976). Splines as linear combinations of B-splines. A survey. University of Wisconsin - Madison Mathematics Research Center Technical Report 1667.

Kaplan, E.L. and Meier, P. (1958). Nonparametric estimation from incomplete observations. Journal of the American Statistical Association **53**, 457-481.

A NONPARAMETRIC ESTIMATOR OF THE SURVIVAL FUNCTION UNDER PROGRESSIVE CENSORING

Joseph C. Gardiner

Michigan State University, East Lansing, Michigan

and

V. Susarla

State University of New York, Binghamton, New York

1. Introduction

The subject of nonparametric estimation of the survival function from incomplete or censored observations has received much attention for more than two decades. We may cite here the celebrated work of Kaplan and Meier (1958) where a product-limit (PL-) estimator of the survival curve is obtained from a sample in which each lifetime may be truncated (fixed censorship) due to limits on observation. In Breslow and Crowley (1974) the properties of this estimator are considered in the case of random censorship, where each lifetime has its own censoring random variable, and the lifetimes and censoring times being each independent and identically distributed (i.i.d.) sequences and also independent of each other. By utilizing the notion of Dirichlet process priors introduced by Ferguson (1973), Susarla and Van Ryzin (1976) obtain a nonparametric Bayesian estimator of the survival function which generalizes the PL-estimator of Kaplan and Meier.

The basic formulation in these works involves consideration of a random sample of lifetimes $X_1,\ldots,X_n$ which may not be completely observable due to the existence of corresponding censoring variables $Y_1,\ldots,Y_n$. The recorded data for the sample is therefore $(Z_1,\delta_1),\ldots,(Z_n,\delta_n)$ where $Z_i = \min(X_i,Y_i)$ and $\delta_i = 0$ or 1 according as $X_i > Y_i$ or $X_i \leq Y_i$. In several longitudinal investigations the variables $\{Z_i : 1 \leq i \leq n\}$ are time-ordered: $Z_{(1)} \leq Z_{(2)} \leq \cdots \leq Z_{(n)}$. The first observation $Z_{(1)}$ is the smallest one followed by the next smallest $Z_{(2)}$ and so on until the largest observation $Z_{(n)}$ is recorded last. In these circumstances cost and time limitations often preclude prolonged experimentation until the complete set of data $\{(Z_i,\delta_i): 1 \leq i \leq n\}$ has been recorded. Furthermore, cogent ethical reasons make it imperative that observation be ceased at the earliest possible stage if the current accumulated data warrants a clear statistical decision. Thus a progressively censored scheme may be advocated in which observation is curtailed at an intermediate stage determined by the cumulative statistical information. If the experimentation is terminated at the k_n^{th} stage, where $k_n \varepsilon \{1,\ldots,n\}$ may be a stopping time, then the recorded data are $\{(Z_{(i)},\delta_i^*): 1 \leq i \leq k_n\}$, with $\delta_i^* = 0$ or 1 according as $Z_{(i)}$ is a censoring time or a true lifetime. The only information available on the remaining $n - k_n$ units is that both their censoring and survival times exceed $Z_{(k_n)}$; that is $Z_{(j)} > Z_{(k_n)}$, $j = k_n + 1,\ldots,n$.

In this paper we construct a nonparametric Bayesian estimator, under squared error loss, for the survival function F from the data $\{(Z_{(i)},\delta_i^*): 1 \leq i \leq k_n;\ Z_{(j)} > Z_{(k_n)},\ j = k_n + 1,\ldots,n\}$ when F follows a Dirichlet process prior. Our estimator thus generalizes, to the progressively censored case, the estimator of Susarla and Van Ryzin (1976) and encompasses both fixed and random censorship. It includes, of course, the cases in which the complete sample is observed ($k_n = n$), an extension of an estimator of Ferguson (1973) when no censoring is present and the Kaplan-Meier estimator. It should be noted that in a progressively censored scheme as described here the observed duration variables $\{Z_{(i)}: 1 \leq i \leq k_n\}$ and their corresponding identifiers $\{\delta_i^*: 1 \leq i \leq k_n\}$ are neither independent nor identically distributed. The absence of this important technical facility in the case of progressive censoring (which is available when the complete data set is observed as in the works cited earlier) introduces additional complications and subtleties in the analysis of progressively censored schemes. For some applications of progressive censoring see Sen, et al. (1973, 1978, 1981).

The substantive material in this paper is distributed in the following three sections. Section 2 introduces the basic assumptions, notation and preliminary notions and provides a brief genesis of our estimator. Various special cases are also dealt with here. We have placed the laborious technical manipulations of construction in Section 4, while Section 3 provides a numerical example.

2. Preliminaries

We are concerned with longitudinal studies in which n specimens under test are followed from the onset with either the time to decrement (survival time) X or its competing censoring time Y recorded for each unit up to the time of the kth response, $k \varepsilon \{1,\ldots,n\}$. We suppose the survival distribution F of X is a Dirichlet process (for the definition of a Dirichlet process and other terms, See Ferguson (1973)) and given F, the survival times $X_1,\ldots,X_n$ of the sample are independent and identically distributed (with distribution 1-F). Furthermore, we consider the corresponding censoring times $Y_1,\ldots,Y_n$ to be independent of F, $X_1,\ldots,X_n$, but make no further assumptions on the distribution of the Y_i's themselves.

The objective is to estimate the survival curve

$$F(t) = P[X > t \mid F], \quad t \geq 0 \quad . \tag{1}$$

We do not have at our disposal the complete set of data $\{(Z_i,\delta_i): 1 \leq i \leq n\}$ where

$$Z_i = \min(X_i, Y_i)$$

$$\delta_i = 0 \text{ or } 1 \quad \text{according as } X_i > Y_i \quad \text{or} \quad X_i \leq Y_i \quad ,$$

but rather the first k order statistics $\{Z_{(1)},\ldots,Z_{(k)}\}$ from $\{Z_1,\ldots,Z_n\}$ and their corresponding identifiers $\{\delta_i^*,\ldots,\delta_k^*\}$, where $\delta_i^* = 1$ or 0 according as $Z_{(i)}$ is a survival time or censoring time, as well as the information that $Z_{(j)} > Z_{(k)}$, $j=k+1,\ldots,n$. On the basis of these recorded data we seek the Bayes estimator $\hat{F}(t)$ of $F(t)$ under the loss ℓ

$$\ell(\hat{F},F) = \int_0^\infty (\hat{F}(x) - F(x))^2 \, dw(x) \quad ,$$

where w is a weight function. Thus $\hat{F}(t)$ is simply the posterior conditional expectation of $F(t)$ given the data; that is, we need to evaluate $E(F(t)\,|\,(Z_{(i)},\delta_i^*):$ $1 \le i \le k,\ Z_{(j)} > Z_{(k)},\ j=k+1,\ldots,n)$, where E denotes expectation with respect to the Dirichlet process with parameter α. As argued in Susarla and Van Ryzin (1976) this may be accomplished in two stages. First relabel the data $\{(Z_{(i)},\delta_i^*)\colon 1 \le i \le k\}$ as follows: let $Z^*_{(1)},\ldots,Z^*_{(\ell)}$ and $Z^*_{(\ell+1)},\ldots,Z^*_{(k)}$ denote, respectively, the ordered survival times and ordered censoring times recorded among $Z_{(1)},\ldots,Z_{(k)}$. Now consider a random sample of size ℓ, say $\eta_1,\ldots,\eta_\ell$ from a Dirichlet process ξ with parameter α and then a random sample size $(k-\ell)$, say $\eta_{\ell+1},\ldots,\eta_k$ from the conditional process of ξ given $\eta_1,\ldots,\eta_\ell$. Then this conditional process is itself a Dirichlet process with parameter β, with β given by

$$\beta(\cdot) = \alpha(\cdot) + \sum_{i=1}^{\ell} I_{\{\eta_i\}}(\cdot) \quad .$$

Therefore if the conditional process of F given $(Z^*_{(1)},1),\ldots,(Z^*_{(\ell)},1)$ is a Dirichlet process with parameter

$$\beta = \alpha + \sum_{i=1}^{\ell} I_{\{Z^*_{(i)}\}} \quad , \tag{2}$$

then the construction of our estimator $\hat{F}(t)$ reduces to the evaluation of $E(F(t)\,|\,(Z^*_{(i)},0)\colon \ell < i \le k,\ Z_{(j)} > Z_{(k)},\ j=k+1,\ldots,n)$, where E now denotes expectation with respect to the distribution of the Dirichlet process with parameter β of (2). This will be shown to reduce to

$$(3) \qquad \hat{F}(t) = \frac{E(F(t)\{\prod_{i=\ell+1}^{k} F(Z^*_{(i)})\}\{F(Z_{(k)})\}^{n-k})}{E(\{\prod_{i=\ell+1}^{k} F(Z^*_{(i)})\}\{F(Z_{(k)})\}^{n-k})} .$$

We shall defer the details of the evaluation of (3) to Section 4. The final form of $\hat{F}(t)$ can be written as

$$(4) \qquad \hat{F}(t) = B(t)W(t)$$

where

$$B(t) = \frac{\alpha(t,\infty) + N^+(t) + (n-k)[t < Z_{(k)}]}{\alpha(R^+) + n}$$

$$W(t) = \prod_{j=1}^{k} \left\{ \frac{\alpha(Z_{(j)},\infty) + N^+(Z_{(j)}) + (n-k) + \lambda_j}{\alpha(Z_{(j)},\infty) + N^+(Z_{(j)}) + (n-k)} \right\}^{[Z_{(j)} \leq t,\ \delta^*_j = 0]/\lambda_j}$$

$$\cdot \left\{ \frac{\alpha(Z_{(k)},\infty) + (n-k)}{\alpha(Z_{(k)},\infty)} \right\}^{[Z_{(k)} \leq t]} ,$$

and $[A]$ denotes the indicator of a set A. Also,

$$N^+(t) = \sum_{j=1}^{k} [Z_{(j)} > t] ,$$

$$\alpha(R^+) = \alpha(0,\infty) ,$$

λ_j = number of censored observations tied at $Z_{(j)}$, and $\frac{0}{0}$ in an exponent is interpreted as unity. It is easy to see that $\hat{F}$ is left continuous at censored observations provided the measure α has no atoms at these points.

Several special cases follow from (4):

(a) Suppose the entire data set $\{(Z_i,\delta_i): 1 \le i \le n\}$ is available. Then setting $k = n$ throughout we obtain

$$(5) \qquad \hat{F}(t) = \left\{\frac{\alpha(t,\infty) + N^+(t)}{\alpha(R^+) + n}\right\} \prod_{j=1}^{n} \left\{\frac{\alpha(Z_j,\infty) + N^+(Z_j) + \lambda_j}{\alpha(Z_j,\infty) + N^+(Z_j)}\right\}^{[Z_j \le t,\delta_j = 0]/\lambda_j}$$

which is the estimator given by Susarla and Van Ryzin (1976). It is also shown there that in the limit $\alpha(R^+) \to 0$, (5) reduces to the Kaplan-Meier product-limit estimator. If, however, we have only the partial data set $\{(Z_{(i)},\delta_i^*): 1 \le i \le k,\ Z_{(j)} > Z_{(k)},\ j = k+1,\dots,n\}$ of a progressively censored sample, then for $t < Z_{(k)}$, the limit of (4) as $\alpha(R^+) \to 0$ is

$$(6) \qquad \frac{N^+(t) + (n-k)}{n} \prod_{j=1}^{k} \left\{\frac{N^+(Z_{(j)}) + (n-k) + \lambda_j}{N^+(Z_{(j)}) + (n-k)}\right\}^{[Z_{(j)} \le t,\delta_j^* = 0]/\lambda_j} .$$

Now writing $\dfrac{N^+(t) + (n-k)}{n} = \prod_{\{j:Z_{(j)} \le t\}} \left\{\dfrac{N^+(Z_{(j)}) + (n-k)}{N^+(Z_{(j)}) + (n-k) + \lambda_j^*}\right\}$ where λ_j^* is the multiplicity of $Z_{(j)}$, we find that (6) reduces to

$$(7) \qquad \prod_{j=1}^{k} \left\{\frac{N^+(Z_{(j)}) + (n-k)}{N^{\prime}(Z_{(j)}) + (n-k) + \lambda_j^*}\right\}^{[Z_{(j)} \le t,\delta_j^* = 1]}$$

If there are no ties among the uncensored observations this is precisely the product-limit estimator for $t < Z_{(k)}$. When $t \geq Z_{(k)}$ the behavior of $\hat{F}$ depends on α even in the limit and we cannot recover $\hat{F}$ since for any $M > 0$ one can choose measures α_1, α_2 which agree on $(0,M]$ but differ on (M,∞).

(b) Suppose that in addition to the entire set $\{(Z_i,\delta_i): 1 \leq i \leq n\}$ being available there is no censoring present. Then setting $k = n$ and $[Z_{(j)} \leq t, \delta_i^* = 0] = 0$ in the terms following (4) we obtain

$$\hat{F}(t) = \frac{\alpha(t,\infty) + N^+(t)}{\alpha(R^+) + n} , \tag{8}$$

which is the estimator of $F(t)$ proposed by Ferguson (1973). Again in the limit as $\alpha(R^+) \to 0$, (8) reduces to

$$\hat{F}(t) = \frac{N^+(t)}{n} = n^{-1} \sum_{j=1}^{n} [X_j > t] , \tag{9}$$

which is the empirical survival function of $X_1,\ldots,X_n$.

On the other hand, if under a progressively censored scheme the only available data are $\{(Z_{(i)},\delta_i^*): 1 \leq i \leq k,\ Z_{(j)} > Z_{(k)},\ j = k+1,\ldots,n\}$, then the absence of observed censoring times among $Z_{(1)},\ldots,Z_{(k)}$ reduces (4) to

$$\hat{F}(t) = \left\{\frac{\alpha(t,\infty) + N^+(t) + (n-k)[t < Z_{(k)}]}{\alpha(R^+) + n}\right\} \left\{\frac{\alpha(Z_{(k)},\infty) + (n-k)}{\alpha(Z_{(k)},\infty)}\right\}^{[Z_{(k)} \leq t]} . \tag{10}$$

We may thus regard (10) as the appropriate generalization of the Ferguson estimator (8) to the progressively censored case. Observe that if $t < Z_{(k)}$, the limit of (10) as $\alpha(R^+) \to 0$ is again the empirical survival function (9). For $t \geq Z_{(k)}$ this limit will depend on α and our previous remark in (a) applies.

With the restrictions noted here we depict the interrelation among the various estimators of F(t) diagrammatically in Figure 1:

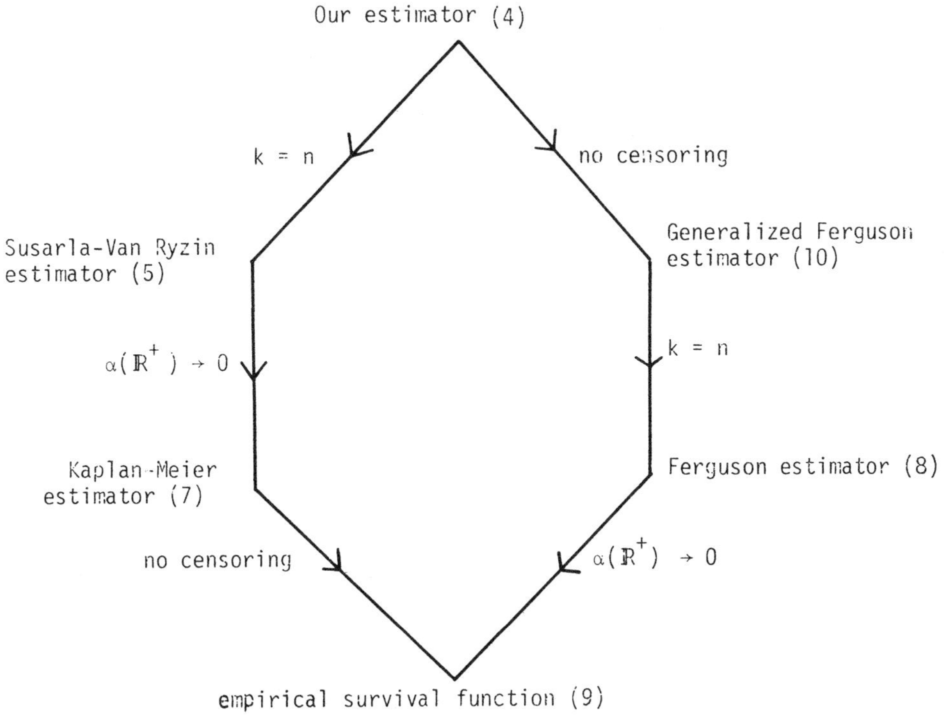

FIGURE 1: Various estimators of F(t)

3. <u>A Numerical Example</u>

We illustrate here the power of a progressively censored scheme with the partial data set $\{(Z_{(i)},\delta_i^*): 1 \leq i \leq k,\ Z_{(j)} > Z_{(k)},\ j = k+1,\ldots,n\}$ $(k<n)$ to yield results that are almost in agreement with those obtained when the complete survival profiles $\{(Z_i,\delta_i): 1 \leq i \leq n\}$ of the sample have been recorded. The data, taken from Johnson and Elandt-Johnson (1980) (p. 179) represent the survival times in weeks of 81 patients in a melanoma study conducted through

the Central Oncology Group at the University of Wisconsin, Madison.

The censored survival times are indicated by a + sign.

136,	58,	55+,	181+,	21,	23,	190+,	65,	234,
194+,	14,	90,	20,	130,	213+,	215+,	124,	108+,
54,	98,	193+,	138,	141,	110,	67+,	50,	26,
103,	59,	134+,	147+,	152+,	65,	40,	34,	57,
81+,	152+,	125+,	151+,	34,	158+,	27,	148+,	27,
132+,	140+,	32,	130+,	38,	85,	129+,	100+,	19,
118,	53,	120+,	66,	46,	37,	50+,	114+,	124+,
26,	102,	93+,	80+,	60,	86+,	21+,	44+,	23,
70,	73+,	19,	38,	31,	25,	76+,	13,	16+,

We choose for our parameter α the measure generated through $\alpha(t,\infty) = \exp(-\theta t)$, $t \geq 0$ where $\theta > 0$ is a real parameter. Since from (1) $E(F(t)) = \alpha(t,\infty)/\alpha(R^+) = e^{-\theta t}$ (expectation with respect to the Dirichlet process F), we estimate θ by

$$\hat{\theta}_k = \sum_{i=1}^{k} \delta_i^* / \sum_{i=1}^{k} Z_{(i)} \quad .$$

A reason for this is that when the censoring times $\{Y_i : 1 \leq i \leq n\}$ are i.i.d. and the survival times are i.i.d. with survival distribution $F_0(t) = \exp(-\theta t)$ then $\{\hat{\theta}_{k_n} : n \geq 1\}$ converges in probability to θ when $n^{-1}k_n \to 1$ as $n \to \infty$. We have computed $\hat{F}(t)$ in three cases: 1) $k = n = 81$; 2) $k = 73$ and 3) $k = 65$. The agreement between the three curves is very good for time points $< Z_{(k)}$.

t =	25	44	54	65	76	100	148	190
1)	.89779	.74449	.69184	.61091	.59691	.54368	.35401	.31245
2)	.89736	.74382	.69107	.61005	.59597	.53361	.35259	.24209
3)	.89691	.74317	.69034	.60925	.59512	.53267	.34307	.22655

4. Proofs

Recall the notation introduced in Section 2. The proof that the conditional process of F given $\{(Z^*_{(i)},1): 1 \leq i \leq \ell\}$ is a Dirichlet process with parameter β as specified in (2), follows along exactly the same lines as that of Theorem 4 in Susarla and Van Ryzin (1976). In order to demonstrate (3) write

$$(11) \qquad E(F(t) \mid (Z^*_{(i)},0): \ell < i \leq k) = \int_0^1 P[F(t) > a \mid (Z^*_{(i)},0): \ell < i \leq k]da \quad ,$$

where E denotes (and in the sequel) expectation with respect to the Dirichlet process with parameter β. Now for $\ell < i \leq k$, $\delta^*_i = 0$, $Z^*_{(i)} = X^*_i \wedge Y^*_i = Y^*_i$ so that $X^*_i > Z^*_{(i)}$ and for $k < j \leq n$ of course X^*_j, $Y^*_j > Z_{(k)}$ where $Z_{(j)} = X^*_j \wedge Y^*_j$. Therefore the integrand in (11) may be written as the ratio I_1/I_2 where

$$(12) \qquad I_1 = E(P[F(t) > a,\ X^*_i \in (Z^*_{(i)},\infty),\ \ell < i \leq k,\ X^*_j,\ Y^*_j \in (Z_{(k)},\infty),\ k < j \leq n \mid F(t),\ F(Z_{(k)}),\ F(Z^*_{(i)}),\ \ell < i \leq k]) \quad ,$$

and I_2 is the resulting expectation obtained by suppressing both F(t) and "$F(t) > a$" in (12). Since $(Y_1,\ldots,Y_n)$ is independent of $(F, X_1,\ldots,X_n)$ and including the terms $Z^*_{(i)} \equiv X^*_i \in (0,\infty)$ (when $\delta^*_i = 1$), $1 \leq i \leq \ell$ we get on simplification

$$(13) \qquad I_1 = E([F(t) > a] \cdot P[X^*_i \in (0,\infty),\ 1 \leq i \leq \ell,\ X^*_i \in (Z^*_{(i)},\infty),\ \ell < i \leq k\ ,\ X^*_j \in (Z_{(k)},\infty),\ k < j \leq n \mid F(t),\ F(Z_{(k)}),\ F(Z^*_{(i)}),\ \ell < i \leq k]) \cdot P[Y^*_j \in (Z_{(k)},\infty),\ k < j \leq n] \quad .$$

Likewise I_2 is obtained from (13) by suppressing $[F(t) > a]$ and $F(t)$. Since $\{X_1,\ldots,X_n\}$ is a random sample from F, the inner conditional probability in (13) is almost surely

$$\{\prod_{i=\ell+1}^{k} F(Z^*_{(i)})\}\{F(Z_{(k)})\}^{n-k} \quad . \tag{14}$$

Finally, using (14) in I_1/I_2 and carrying out the integration in (11) we obtain (3).

We are now left with the tedious task of carrying out the integrations in (3). Several cases must be considered depending on the position of the time point t among the observed points $Z^*_{(\ell+1)},\ldots,Z^*_{(k)}$ and $Z_{(k)}$ $(=\max\{Z_{(i)}: 1 \leq i \leq k\})$.

Suppose $Z^+_{(\ell+1)},\ldots,Z^+_{(m)}$ denote the distinct ordered values among $Z^*_{(\ell+1)},\ldots, Z^*_{(k)}$ with corresponding multiplicities $\lambda_{\ell+1},\ldots,\lambda_m$. Thus

$$\lambda_i \geq 1,\ \ell < i \leq m;\quad 1 \leq \ell \leq k \leq n \quad \text{and} \quad \ell < m \leq k,\ \sum_{i=\ell+1}^{n} \lambda_i = k-\ell \quad .$$

The largest recorded observable $Z_{(k)}$ may be either a survival time or a censoring time. Suppose we are in the latter case so that $Z_{(k)} = Z^+_{(m)}$. Consider the case $t > Z^+_{(m)}$. Select the partition of $R^+ = (0,\infty)$ given by the points $\{Z^+_{(i)}: \ell \leq i \leq m+2\}$ where $Z^+_{(\ell)} = 0, Z^+_{(m+1)} = t$ and $Z^+_{(m+2)} = \infty$. Then defining $U_i = F(Z^+_{(i)}) - F(Z^+_{(i+1)}), \ell < i \leq m+1$, the random vector $(U_\ell,\ldots,U_{m+1})$ has the Dirichlet distribution with parameter $(\beta_\ell,\ldots,\beta_{m+1})$ where

$$\beta_i = \beta(Z^+_{(i)}, Z^+_{(i+1)}] \tag{15}$$

and β as given in (2).

Now $F(Z^+_{(i)}) = (1 - \sum_{j=\ell}^{i-1} U_j)$, $\ell < i \leq m+1$. Therefore the integrand in the numerator of (3), $F(t)\{\prod_{i=\ell+1}^{m} (F(Z^+_{(i)}))^{\lambda_i}\}\{F(Z^+_{(m)})^{n-k}\}$, can be written

$$(16) \qquad \prod_{i=\ell+1}^{m+1} \left(1 - \sum_{j=\ell}^{i-1} U_j\right)^{\lambda_i} \left(1 - \sum_{j=\ell}^{m-1} U_j\right)^{n-k} ,$$

with $\lambda_{m+1} = 1$. For the denominator in (3) the integrand is the same as (16) except that $\lambda_{m+1} = 0$.

Now $(U_\ell, \ldots, U_m)$ has probability density

$$\left\{\frac{\Gamma\left(\sum_{i=\ell}^{m+1} \beta_i\right)}{\prod_{i=\ell}^{m+1} \Gamma(\beta_i)}\right\} \left\{\prod_{j=\ell}^{m} u_j^{\beta_j - 1}\right\} \left(1 - \sum_{j=\ell}^{m} u_j\right)^{\beta_{m+1} - 1} ,$$

$$0 < \text{all } u_j < 1$$

$$\text{and} \quad 0 < \sum_{j=\ell}^{m} u_j < 1 ,$$

where Γ denotes the Gamma function. The expectation of (16) involves integration over the variables $u_\ell, \ldots, u_m$. Suppressing all terms not involving u_m, the integral over u_m is

$$\int_0^{(1 - \sum_{j=\ell}^{m-1} u_j)} u_m^{\beta_m - 1} \left(1 - \sum_{j=\ell}^{m-1} u_j - u_m\right)^{(\beta_{m+1} + \lambda_{m+1}) - 1} du_m$$

$$= \frac{\Gamma(\beta_m)\, \Gamma(\beta_{m+1} + \lambda_{m+1})}{\Gamma(\beta_m + \beta_{m+1} + \lambda_{m+1}} \left(1 - \sum_{j=\ell}^{m-1} u_j\right)^{\beta_m + \beta_{m+1} + \lambda_{m+1} - 1} .$$

Now proceeding with the successive integrations over $u_{m-1}, \ldots, u_\ell$ we finally obtain for the numerator of (3)

$$(17)\quad \left\{\frac{\Gamma\left(\sum_{i=\ell}^{m+1}\beta_i\right)}{\prod_{i=\ell}^{m+1}\Gamma(\beta_i)}\right\}\left\{\frac{\Gamma(\beta_m)\,\Gamma(\beta_{m+1}+\lambda_{m+1})}{\Gamma(\beta_m+\beta_{m+1}+\lambda_{m+1})}\right\}\prod_{i=\ell}^{m-1}\left\{\frac{\Gamma(\beta_i)\,\Gamma\left(\sum_{j=i+1}^{m+1}(\beta_j+\lambda_j)+(n-k)\right)}{\Gamma\left(\beta_i+\sum_{j=i+1}^{m+1}(\beta_j+\lambda_j)+(n-k)\right)}\right\}.$$

Note that in (17) the value of λ_{m+1} is 1. For the denominator of (3) we have the same expression except that $\lambda_{m+1}=0$.

Proceeding with cancellations of the common factors in the numerator and denominator of (3) yields our estimator

$$(18)\qquad \hat{F}(t)=\left(\frac{\beta_{m+1}}{\beta_m+\beta_{m+1}}\right)\prod_{j=\ell+1}^{m}\left\{\frac{\sum_{i=j}^{m}(\beta_i+\lambda_i)+\beta_{m+1}+(n-k)}{\sum_{i=j}^{m}(\beta_i+\lambda_i)+\beta_{m+1}+(n-k)+\beta_{j-1}}\right\}.$$

Recall (15) and (2). A trite computation shows

$$\beta_{m+1}=\alpha(t,\infty)+\#\,(\text{observed lifetimes} > t)$$

$$\beta_m+\beta_{m+1}=\alpha(Z^+_{(m)},\infty)+\#\,(\text{observed lifetimes} > Z^+_{(m)})$$

$$(19)\quad \sum_{i=j}^{m}(\beta_i+\lambda_i)+\beta_{m+1}=\alpha(Z^+_{(j)},\infty)+\#\,(\text{observed lifetimes} > Z^+_{(j)})$$

$$+\#\,(\text{observed censoring times} \geq Z^+_{(j)})$$

$$=\alpha(Z^+_{(j)},\infty)+N^+(Z^+_{(j)})+\lambda_j$$

$$\sum_{i=j}^{m}(\beta_i+\lambda_i)+\beta_{m+1}+\beta_{j-1}=\alpha(Z^+_{(j-1)},\infty)+N^+(Z^+_{(j-1)})\quad .$$

Substituting in (18) and again cancelling out common factors we get for the product term in (18)

$$(20)\qquad \frac{\alpha(Z^+_{(m)},\infty) + N^+(Z^+_{(m)}) + (n-k) + \lambda_m}{\alpha(Z^+_{(\ell)},\infty) + N^+(Z^+_{(\ell)}) + (n-k)} \prod_{j=\ell+1}^{m} \left\{ \frac{\alpha(Z^+_{(j)},\infty) + N^+(Z^+_{(j)}) + (n-k)+\lambda_j}{\alpha(Z^+_{(j)},\infty) + N^+(Z^+_{(j)}) + (n-k)} \right\}.$$

Now $Z^+_{(m)} = Z_{(k)}$ and $Z^+_{(\ell)} = 0$. Also in the case considered here $N^+(Z^+_{(m)}) = 0$, $N^+(t) = 0$. Using these in (19) and (20) yields

$$\hat{F}(t) = \frac{\alpha(t,\infty)}{\alpha(R^+) + n} \prod_{j=\ell+1}^{m} \left\{ \frac{\alpha(Z^+_{(j)},\infty) + N^+(Z^+_{(j)}) + (n-k) + \lambda_j}{\alpha(Z^+_{(j)},\infty) + N^+(Z^+_{(j)}) + (n-k)} \right\}$$

$$\cdot \left\{ \frac{\alpha(Z_{(k)},\infty) + (n-k)}{\alpha(Z_{(k)},\infty)} \right\}$$

which is the form of (4) for this case.

All other cases are handled in exactly the same manner and lead to the general form of $\hat{F}(t)$ given in (4).

5. Concluding Remarks

It can be shown that when the censoring times $\{Y_i: i \geq 1\}$ are i.i.d. with continuous right distribution function G on $(0,\infty)$, the survival times $\{X_i: i \geq 1\}$ are i.i.d. with continuous right distribution F, and $n^{-1}k_n \to \gamma \in (0,1]$, then for any $T > 0$ with $F(T)G(T) > 1-\gamma$, the process $\{n^{\frac{1}{2}}(\hat{F}(t) - F(t)): t \in [0,T\}$ converges weakly to a Gaussian process. Furthermore under appropriate conditions strong convergence and consistency can be demonstrated.

ACKNOWLEDGEMENTS

The authors wish to thank Professor P.K. Wong for his programming assistance. Research sponsored in part by the National Institutes of Health under Grants NIGMS-1R01-GM28030 and NIGMS-1R01-GM28405, and in part by the Office of Naval Research under contract N00014-79-C0522.

REFERENCES

Breslow, N. and Crowley, J. (1974). A large sample study of the life table and product limit estimates under random censorship. Annals of Statistics 2, 437-453.

Chatterjee, S.K. and Sen, P.K. (1973). Nonparametric testing under progressive censoring. Calcutta Statistical Association Bulletin 22, 13-50.

Delong, E. and Sen, P.K. (1981). Estimation of P[X > Y] based on progressively truncated versions of the Wilcoxon-Mann-Whitney statistics. Communications in Statistics A10 (10), 963-981.

Ferguson, T.S. (1973). A Bayesian analysis of some nonparametric problems. Annals of Statistics 1, 209-230.

Gardiner, J. and Sen, P.K. (1978). Asymptotic normality of a class of time-sequential statistics and applications. Communications in Statistics A7 4, 373-388.

Johnson, N. and Elandt-Johnson, R. (1980). Survival Models and Data Analysis. Wiley & Sons, New York.

Kaplan, E.L. and Meier, P. (1958). Nonparametric estimation from incomplete observations. Journal of the American Statistical Association 53, 457-481.

Susarla, V. and Van Ryzin, J. (1976). Nonparametric Bayesian estimation of survival curves from incomplete observations. Journal of the American Statistical Association 61, 897-902.

FOURIER INTEGRAL ESTIMATE OF THE FAILURE RATE FUNCTION AND ITS MEAN SQUARE ERROR PROPERTIES

Nozer D. Singpurwalla

The George Washington University, Washington, D.C.

and

Man-Yuen Wong

Automated Sciences Group, Inc., Silver Spring, Maryland

1. Introduction

The failure rate function h is important in reliability and biometry. Estimates of h using weighting functions or "kernels" are quite common in the literature (see Singpurwalla and Wong (1982b)). The kernels that have been considered so far are *nonnegative* and *absolutely integrable in* $(-\infty,\infty)$. (Kernels satisfying this latter condition are known as L^1 kernels.) Singpurwalla and Wong (1982a) -- abbreviated as SW (1982a) -- have shown that the mean square error (MSE) of a kernel estimator of h using a compact L^1 kernel restricted to be nonnegative has an optimal rate of convergence of at most $O(n^{-4/5})$, regardless of the smoothness of h; n is the sample size. If the nonnegativity condition of the compact L^1 kernel is relaxed, and if h is $(m+1)$ times continuously differentiable, then (for $m > 2$), the rate of convergence of the MSE (can be improved and) is at most $O(n^{-2m/(2m+1)})$. A method for producing kernel estimators having the above property is the generalized jackknife of Gray and Schucany (1972). Specifically, if we use the generalized jackknife on two kernel estimators of h, with each estimator being based upon a nonnegative compact L^1 kernel, then this is equivalent to directly producing a kernel estimator of h using a compact L^1 kernel which takes both positive and negative values. If we continue to apply the generalized jackknife method, then the rate of convergence of the MSE of the resulting estimator can be brought as close to n^{-1} as is desired. This, plus the alternating behavior of the resulting kernel, has prompted us to conjecture that a repeated jackknifing of estimators based on compact L^1 kernels is equivalent to obtaining a kernel estimator using an alternating (wave-like) non L^1 kernel.

Motivated by the above considerations, our goal in this paper is to obtain an estimator of h whose MSE converges to 0 faster than $O(n^{-2m/(2m+1)})$ for any finite $m > 0$, and preferably is closer to the ideal n^{-1}. We achieve this goal by considering a kernel estimator of h based on the "sinc" kernel. In Section 3 we show that the sinc kernel, which is not an L^1 kernel and may not be a limiting case of jackknifing an L^1 kernel either, arises naturally when we estimate h via an estimate of the Fourier transform of h. The sinc kernel estimator of h is also referred to as the "Fourier integral estimate".

In Section 4 we show that the sinc kernel estimators of h are asymptotically unbiased and consistent. In Section 5, we discuss the rates of convergence of the bias and the MSE of these estimators. We show that for certain classes of failure rate functions, the sinc kernel estimators have a faster rate of convergence of the MSE than the corresponding L^1 kernel estimators. These rates are of the order $(\log n/n)$ or $(n^{(1/(2p-1))-1})$, depending upon whether the Fourier transform of h decreases "exponentially" or "algebraically with degree p" (see Definitions 5.1 and 5.2). Clearly, when $p > m+1$, both the above rates are faster than $n^{-2m/(2m+1)}$.

Sinc kernels have been considered before in the literature, first by Konakov (1972), and more recently by Davis (1975) on density estimation. Thus, the results of our paper complement those of Konakov and Davis.

2. Preliminaries: Kernel Estimates

Suppose that the time to failure of a device is a nonnegative random variable X, with an absolutely continuous distribution function F and a probability density function f. The *failure rate* at x_0, $h(x_0)$, for $F(x_0) \neq 1$, is defined as

$$h(x_0) = \frac{f(x_0)}{1 - F(x_0)} \; ;$$

note that $h(x) \geq 0$, for all $x \geq 0$.

Given an ordered sample of n lifetimes from F, say $X_{(1)}, \ldots, X_{(n)}$, a *kernel estimate* of $h(x_0)$, $h(n,x_0)$, is defined as

$$(1) \qquad h(n,x_0) = \sum_{j=1}^{n} \frac{1}{n-j+1} \frac{1}{b(n)} K\left(\frac{X_{(j)} - x_0}{b(n)}\right) ,$$

where the kernel K is a bounded, symmetric function of integral one; the scale parameter b(n) is a nonnegative decreasing function of n such that

$$(2) \qquad \text{(i)} \;\; \lim b(n) = 0, \text{ and } n \to \infty , \qquad \text{(ii)} \;\; \lim n\, b(n) = \infty \cdot n \to \infty .$$

A motivation for considering the kernel estimates of the failure rate are given in Watson and Leadbetter (1964a).

Watson and Leadbetter (1964b) have shown that for a certain class of distribution functions, estimates based on L^1 kernels are asymptotically unbiased and consistent, at every point x at which h is continuous and $F(x) < 1$. The optimal rates of convergence of the bias and the MSE of $h(n,x_0)$ have been discussed by SW (1982b).

3. Kernel Estimates Based on the Fourier Integral

We shall confine our attention to the class of failure rate functions h for which the Fourier transform Φ_h exists; that is

$$\Phi_h(x) = \int e^{ixu} h(u)du < \infty .$$

Let x_0 be a point of continuity of h(x), and assuming that $\Phi_h \in L^1$ (i.e., $\int |\Phi_h(x)| dx < \infty$, the following inversion formula gives us the basis for considering the Fourier integral estimate of the failure rate:

$$h(x_0) = \frac{1}{2\pi} \int e^{-ix_0 u} \Phi_h(u)du . \tag{3}$$

Let F_n be the *modified sample distribution function*; that is, the usual sample distribution function multiplied by n/(n+1). An estimate of h(x) at $x = X_{(j)}$, $h_n(x)$, is

$$h_n(x) = \frac{f_n(x)}{1 - F_n(x)} = \frac{dF_n(x)}{1 - F_n(x)} = \frac{1}{n - j+1} .$$

Let Φ_{h_n} be the Fourier transform of h_n; that is

$$\Phi_{h_n}(x) = \int e^{ixu} h_n(u)du = \sum_{j=1}^{n} \frac{1}{n - j+1} e^{ixX_{(j)}} . \tag{4}$$

To obtain from (3) an estimate of $h(x_0)$, we replace Φ_h by Φ_{h_n}, and to assure finiteness of the integral, we take it between the finite limits $(-\frac{1}{b(n)}, \frac{1}{b(n)})$, where the b(n) satisfy (2), we obtain the *Fourier integral estimator* of $h(x_0)$, $\tilde{h}(n,x_0)$, where

$$(5)\qquad \tilde{h}(n,x_0) = \frac{1}{2\pi}\int_{-\frac{1}{b(n)}}^{\frac{1}{b(n)}} e^{-ix_0 u}\,\Phi_{h_n}(u)\,du \quad .$$

A simple computation shows that

$$(6)\qquad \tilde{h}(n,x_0) = \sum_{j=1}^{n} \frac{1}{(n-j+1)b(n)}\, S\left(\frac{X_{(j)}-x_0}{b(n)}\right) ,$$

where $S(x) = (\sin x)/\pi x$ is the "sinc" function.

Thus we see that the Fourier integral estimate of the failure rate is indeed a kernel estimate, with the sinc function S as the kernel.

Note that the kernel S is not an L^1 kernel, but that it is symmetric, bounded, and of integral one; also $\int S^2(x)dx = 1/\pi$.

4. Asymptotic Unbiasedness and Consistency

Since S is not an L^1 kernel, the asymptotic unbiasedness and consistency of $\tilde{h}(n,x_0)$ has to be established first. Once this is done, we will be able to discuss the rates of convergence of the bias and the MSE.

THEOREM 1: Let $X_{(1)} \leq X_{(2)} \leq \cdots \leq X_{(n)}$ be an ordered sample of lifetimes from an absolutely continuous distribution function F. Suppose that:

(i) the failure rate function h is absolutely integrable;

(ii) h satisfies Dirichlet's conditions in any finite interval; that is, h has at most a finite number of finite discontinuities, and no infinite discontinuities in any finite interval, and, furthermore, h has only a finite number of maxima and minima in any finite interval;

(iii) $h(x)$ is continuous at x_0;

(iv) $F(x_0) < 1$; and

(v) F is such that for any fixed x', and every fixed $\lambda > 0$, there exists a $G_\lambda > 0$, such that

$$\frac{1}{\pi(1-F(x))} \left| \frac{\sin((x-x')/b(n))}{x-x'} \right| \leqq G_\lambda$$

for all sufficiently large n and for all

$$|x - x'| \geqq \lambda ,$$

then $\tilde{h}(n,x_0)$ defined by (6) is an asymptotically unbiased and consistent estimator of $h(x_0)$.

Furthermore, an asymptotic expression for the expected value of $\tilde{h}(n,x_0)$ is*

(7) $$E[\tilde{h}(n,x_0)] \sim \int \frac{1}{\pi} \frac{\sin((u-x_0)/b(n))}{u-x_0} h(u)du ,$$

and the variance $Var[\tilde{h}(n,x_0)]$ converges to zero at the rate $1/nb(n)$.

PROOF: $E[\tilde{h}(n,x_0)]$

(8) $$= \sum_{j=1}^{n} \int \frac{1}{n-j+1} \frac{\sin((u-x_0)/b(n))}{\pi(u-x_0)} f_{X_{(j)}}(u)du$$

$$= \frac{1}{\pi} \int \frac{\sin((u-x_0)/b(n))}{u-x_0} h(u)du - \frac{1}{\pi} \int \frac{\sin((u-x_0)/b(n))}{u-x_0} h(u) F^n(u)du .$$

*The notation "$a_n \sim b_n$" denotes the fact that the ratio of a_n to b_n has limit one.

Consider the limit of the first term on the right-hand side of (8):

$$(9)\qquad \lim_{n\to\infty} \frac{1}{\pi}\int \frac{\sin((u-x_0)/b(n))}{u-x_0}\, h(u)du = \lim_{b(n)\to 0} \frac{1}{\pi}\int \frac{\sin((u-x_0)/b(n))}{u-x_0}\, h(u)du$$

$$= h(x_0) \quad .$$

The last equality follows by the Fourier integral formula (see Titchmarsh (1962, pp. 3,25)).

Next we show that the second term on the right-hand side of (8) tends to zero, as $n \to \infty$. Since $F(x_0) < 1$, we can choose a $\lambda > 0$ so that $F(x_0 + \lambda) < 1$, and such that $h(u)$ is bounded in $|u-x_0| \leq \lambda$. We split the interval of integration $(-\infty,\infty)$ into two parts, $|u-x_0| \leq \lambda$ and $|u-x_0| > \lambda$, and note that

$$(10)\qquad \int_{|u-x_0| \leq \lambda} \frac{1}{\pi} \frac{\sin\left(\frac{u-x_0}{b(n)}\right)}{u-x_0}\, h(u)\, F^n(u)du \leq (\text{const})\ F^n(x_0+\lambda) \to 0 \ ,$$

as $n \to \infty$, and

$$(11)\qquad \int_{|u-x_0| > \lambda} \frac{1}{\pi} \frac{\sin\left(\frac{u-x_0}{b(n)}\right)}{u-x_0}\, h(u)\, F^n(u)du \leq G_\lambda \int_0^1 F^n dF = \frac{G_\lambda}{n+1} \to 0 \ ,$$

as $n \to \infty$.

From (8) through (11), we conclude that

$$E[\tilde{h}(n,x_0)] \sim \int \frac{1}{\pi} \frac{\sin\left(\frac{u-x_0}{b(n)}\right)}{u-x_0}\, h(u)du \to h(x_0) \text{ as } n \to \infty \quad .$$

To prove consistency of $\tilde{h}(n,x_0)$, we follow the detailed steps given in Watson and Leadbetter (1964b). From equation (4) of Watson and Leadbetter, we write

$$\mathrm{Var}[\tilde{h}(n,x_0)]$$

$$= \int \frac{1}{b^2(n)} S^2\left(\frac{u-x_0}{b(n)}\right) h(u)\, I_n(F(u))\, dF(u)$$

(12) $$+ 2\iint\limits_{0\le u\le v} \frac{\frac{1}{b(n)} S\left(\frac{u-x_0}{b(n)}\right) \frac{1}{b(n)} S\left(\frac{v-x_0}{b(n)}\right)}{1-F(v)} \left\{\frac{1-F^n(u)}{1-F(u)} F^n(v) - \frac{F^n(v)-F^n(u)}{F(v)-F(u)}\right\} dF(u)\, dF(v) ,$$

where $I_n(F) = \int_0^{1-F} \frac{(F+B)^n - F^n}{B}\, dB$.

If we multiply both sides of (12) by n/α_n, where

$$\alpha_n = \int \frac{1}{b^2(n)} S^2\left(\frac{u-x_0}{b(n)}\right) du = \frac{1}{\pi b(n)}$$

and take the limit as $n \to \infty$, we note that the first term on the right-hand side of (12) equals $h(x_0)/(1-F(x_0))$ whereas the second term is 0. Thus

$$\lim_{n\to\infty} \frac{n}{\alpha_n} \mathrm{Var}[\tilde{h}(n,x_0)] = \frac{h(x_0)}{1-F(x_0)}$$

or that

$$\mathrm{Var}[\tilde{h}(n,x_0)] \sim \frac{\alpha_n}{n} \frac{h(x_0)}{1-F(x_0)} .$$

Since $\alpha_n/n = (1/\pi)(1/nb(n)) \to 0$ by (2), it follows that $\mathrm{Var}[\tilde{h}(n,x_0)] \to 0$. Thus, $\tilde{h}(n,x_0)$ is a consistent estimator of $h(x_0)$, and the variance of $\tilde{h}(n,x_0)$ goes to zero at the rate $1/nb(n)$.

4.1 An Alternate Expression for the Bias

We shall find it useful to express the asymptotic bias of $\tilde{h}(n,x_0)$ in terms of the Fourier transform Φ_h of h. We first note that if w(t) is the indicator of the interval [-1,1], then

$$S(x) = \frac{1}{2\pi} \int e^{-ixu}\, w(u)du \quad . \tag{13}$$

In view of the above, the Fourier transform of S(x) is w(x), $|x| \neq 1$. Recall, from (7), that

$$E[\tilde{h}(n,x_0)] \sim \frac{1}{b(n)} \int h(u)\; S((x_0-u)/b(n))du$$

$$= \frac{1}{2\pi} \int e^{-ix_0t}\, w(b(n)t)\; \phi_h(t)dt \quad .$$

The asymptotic bias of $\tilde{h}(n,x_0)$ is therefore given by

$$\mathrm{Bias}[\tilde{h}(n,x_0)] \sim \frac{1}{2\pi} \int e^{-ix_0t}\, w(tb(n))\; \phi_h(t)\; dt - h(x_0)$$

$$= \frac{1}{2\pi} \int e^{-ix_0t}\, \{w(tb(n)) - 1\}\; \phi_h(t)\; dt \tag{14}$$

$$= -\frac{1}{2\pi} \int_{|t| > \frac{1}{b(n)}} e^{-ix_0t}\; \phi_h(t)\; dt \quad .$$

5. Rates of Convergence of the Bias and the MSE

We are able to investigate and optimize the rates of convergence of the bias and the MSE of $\tilde{h}(n,x_0)$ when the Fourier transform of the failure rate decreases exponentially or algebraically.

DEFINITION 1: (Parzen (1962)): A function $g(x)$ is said to *decrease exponentially* with degree $0 < r \leq 2$, and coefficient $\rho > 0$, if

$$|g(x)| \leq Ae^{-\rho|x|^r} \quad \text{for some constant } A > 0 \quad , \tag{15}$$

and

$$\lim_{x\to\infty} \int_0^1 [1+\exp(2\rho x^r)\ |g(xu)|^2]^{-1}\ du = 0\ . \tag{16}$$

We shall first need to prove the following lemmas. The first is a simple application of L'Hôpital's rule.

LEMMA 1:

$$\lim_{n\to\infty} b(n)\ e^{\frac{1}{b^r(n)}} \int_{\frac{1}{b(n)}}^{\infty} e^{-t^r}\ dt = 0,\ (r > 0). \tag{17}$$

LEMMA 2: If the Fourier transform of h, ϕ_h, decreases exponentially with degree r and coefficient ρ, then, for sufficiently large n,

$$|\text{Bias}[\tilde{h}(n,x_0)]| \leq \frac{1}{\pi} \int_{-\frac{1}{b(n)}} Ae^{-\rho t^r}\ dt\ . \tag{18}$$

PROOF: From (14), we note that for n large

$$\text{Bias}[\tilde{h}(n,x_0)] \sim \left| -\frac{1}{2\pi} \int_{|t| > \frac{1}{b(n)}} e^{-ix_0 t} \phi_h(t)dt \right|$$

$$\leq \frac{1}{2\pi} \int_{|t| > \frac{1}{b(n)}} Ae^{-\rho|t|^r} dt = \frac{1}{\pi} \int_{t > \frac{1}{b(n)}} Ae^{-\rho t^r} dt;$$

the statement of the lemma now follows.

LEMMA 3: Suppose that the Fourier transform ϕ_h of h decreases exponentially with degree r and coefficient ρ. Then

$$\lim_{n \to \infty} b(n)\, e^{\rho/b^r(n)} \left|\text{Bias}[\tilde{h}(n,x_0)]\right| = 0 . \tag{19}$$

PROOF: The result follows if we make a change of variable $u^r = \rho t^r$, and use Lemmas 1 and 2.

The following theorem establishes the choice of b(n) which enables us to obtain the optimal rate of convergence of the mean square error of $\tilde{h}(n,x_0)$, when ϕ_h decreases exponentially. It follows from Lemma 3, Theorem 1, and Davis (1975).

THEOREM 2: Suppose that the Fourier transform ϕ_h of the unknown failure rate h exists and decreases exponentially with degree $0 < r \leq 2$ and coefficient $\rho > 0$. Then, if b(n) in the Fourier integral estimator of h, $\tilde{h}(n,x_0)$, given by (6), is chosen such that $b(n) = O(\log n/2\rho)^{-(1/r)}$, the optimal rate of convergence of the MSE of $\tilde{h}(n,x_0)$ is of the order log n/n.

We shall now consider the class of failure rate functions h whose Fourier transforms ϕ_h decrease albegraically.

DEFINITION 2: (Parzen (1962)): A function g(x) is said to *decrease algebraically* with degree $p > 0$, if

(20) $$\lim_{|x|\to\infty} |x|^p \, |g(x)| = \alpha^{\frac{1}{2}} > 0, \text{ for some } \alpha > 0 \; .$$

LEMMA 4: Suppose that the Fourier transform ϕ_h of h decreases algebraically with degree $p > 1$. Then

(21) $$\lim_{n\to\infty} b^{1-p}(n) \int_{|t| > \frac{1}{b(n)}} |\phi_h(t)|\,dt = 2\alpha^{\frac{1}{2}}(p-1)^{-1} \; .$$

PROOF: From (20), we note that for $\varepsilon > o$, there exists an $M > 0$ such that for $|t| > M$

$$|t|^{-p}(\alpha^{\frac{1}{2}} - \varepsilon) < |\phi_h(t)| < |t|^{-p}\,(\alpha^{\frac{1}{2}} + \varepsilon) \; .$$

The proof is completed by integrating both sides of the above for $|t| > 1/b(n)$, and noting that when n is sufficiently large, $1/b(n) > M$, and $|t| > 1/b(n)$ implies that $|t| > M$.

LEMMA 5: Suppose that the Fourier transform ϕ_h of h decreases algebraically with degree $p > 1$. Then the bias of $\tilde{h}(n,x_0)$, $\text{Bias}[\tilde{h}(n,x_0)]$, satisfies

(22) $$\lim_{n\to\infty} b^{1-p}(n) \; |\text{Bias}[\tilde{h}(n,x_0)]| \leqq \alpha^{\frac{1}{2}} \pi^{-1} (p-1)^{-1} \; ;$$

thus the bias decreases at the rate $b^{p-1}(n)$.

PROOF: From (14), we have

$$\left| \frac{1}{2\pi} \int_{|t| > \frac{1}{b(n)}} e^{-ix_0 t} \phi_h(t)dt \right| \leqq \frac{1}{2\pi} \int_{|t| > \frac{1}{b(n)}} |\phi_h(t)|dt \ .$$

We obtain from (14) and (21)

$$\lim_{n\to\infty} b^{1-p}(n)|\mathrm{Bias}[\tilde{h},(n,x_0)]| \leqq \frac{1}{2\pi} 2\alpha^{\frac{1}{2}}(p-1)^{-1} = \alpha^{\frac{1}{2}}\pi^{-1} (p-1)^{-1} \ .$$

The following theorem is analogous to Theorem 2.

THEOREM 3: Suppose that the Fourier transform ϕ_h of the unknown failure rate h exists and decreases algebraically with degree $p > 1$. Then, if $b(n)$ in the Fourier integral estimator of h, $\tilde{h}(n,x_0)$, given by (6), is chosen such that $b(n) = O(n^{-1/(2p-1)})$, the optimal rate of convergence of the MSE of $\tilde{h}(n,x_0)$ is of the order $n^{1/(2p-1)-1}$.

6. A Comparison of the Rates of Convergence of the MSE's

We can now compare the optimal rates of convergence of the MSE's for estimates of h based on L^1 kernels and the sinc function kernel which is not an L^1 kernel.

In general, for L^1 kernels which belong to the class A_m (i.e., an L^1 kernel K which satisfies the condition that $\int x^j K(x)dx = 0$, for $j = 1,2,\ldots,m-1$), and if $h^{(m+1)}$ exists (that is, if h is m times continuously differentiable), then we have shown in SW (1982a) that the optimal rate of convergence of the MSE of the kernel estimator of h is of the order $n^{-2m/(2m+1)}$. The following results are immediate consequences of this and Theorems 2 and 3 of this paper.

THEOREM 4: For the class of failure rate functions whose Fourier transform exists, and decreases exponentially, the Fourier integral estimate (based on the sinc function) is better in terms of the rate of convergence of the MSE than a kernel estimate based on any L^1 kernel.

THEOREM 5: For the class of failure rate functions whose Fourier transform exists and decreases albegraically with degree p, the Fourier integral estimate (based on the sinc function) is better in terms of the MSE than a kernel estimate based on any L^1 kernel belonging to the class A_m, if $p > m+1$.

7. Concluding Remarks

It is evident from Theorems 2 and 3 that one would consider using the Fourier integral estimate of h only when one had some prior knowledge about h. A disadvantage of the sinc kernel estimator stems from the fact that the estimator of h can be negative at some points, a result unacceptable to practitioners. One may argue that this is the price that must be paid for obtaining an estimator which has good bias and MSE properties. On the other hand, a Bayesian may view this as another situation wherein unbiased estimation and MSE minimization lead us to unacceptable answers.

ACKNOWLEDGEMENTS

The work reported here was begun when the authors were visitors at the Department of Statistics of Stanford University (1978-1979). It has been supported by Contract N00014-77-C-0263, Project NR-042-372, Office of Naval Research, and Grant DAAG-29-80-C-0067, U.S. Army Research Office, Durham, North Carolina.

REFERENCES

Chung, K.L. (1974). A course in probability theory. Academic Press, New York.

Davis, K.B. (1975). Mean square error properties of density estimates. Annals of Statistics 3, 1025-1030.

Gray, H.L. and Schucany, W.R. (1972). The generalized jackknife statistics. Marcel Dekker, New York.

Konakov, V.D. (1972). Nonparametric estimation of density functions. Theoretical Probability and Applications 17, 361-362.

Miller, R.G. (1978). The jackknife: survey and applications. Proceedings of the 23rd Conference on the Design of Experiments in Army Research Development and Testing, ARO Report 78-2.

Parzen, E. (1962). On estimation of a probability density function and mode. Annals of Mathematical Statistics 33, 1065-1076.

Rice, J. and Rosenblatt, M. (1976). Estimation of the log survivor function and hazard function. Sankhya A 38, 60-78.

Singpurwalla, N.D. and Wong, M.-Y. (1982a). Improvement of kernel estimators of the failure rate function using the generalized jackknife. Tentative acceptance, Journal of the American Statistical Association.

Singpurwalla, N.D. and Wong, M.-Y. (1982b). Estimation of the failure rate - a survey of nonparametric methods, Part I: Non-Bayesian methods. Communications in Statistics, Statistical Reviews (to appear).

Tirchmarsh, E.C. (1962). Introduction to the theory of Fourier integrals. Oxford University Press, London.

Watson, G.S. and Leadbetter M.R. (1964a). Hazard analysis I. Biometrika 51, 175-184.

Watson, G.S. and Leadbetter, M.R. (1964b). Hazard analysis II. Sankhya A 26, 110-116.

ESTIMATION OF THE RATIO OF HAZARD FUNCTIONS

John Crowley

University of Washington and Fred Hutchinson
Cancer Research Center

P.Y. Liu

University of California at Los Angeles

Joseph G. Voelkel

Allied Chemical Corporation

1. Introduction

In a clinical trial in which comparison of survival across treatment groups is of interest, it is useful to have a descriptive measure of the difference in survival between groups. If the hazard functions in two groups are roughly proportional, then the ratio of hazard functions has the interpretation of relative risk and has intuitive appeal as a descriptive statistic.

In this paper we investigate the large sample properties of several estimators of relative risk. The experimental setting envisioned is the clinical trial or other similar situations in which survival is being measured, and in which there are possibly different potential follow-up times for each patient.

The notation and model are given in Section 2. Section 3 presents the maximum likelihood estimator based on an exponential model, and the maximum partial likelihood estimator of Cox (1972;1975). Various other approaches are indicated in Section 4, and one, the Mantel-Haenszel (1959) estimator, is investigated in detail. Some numerical comparisons are made in Section 5.

2. Notation

For the ith sample, $i=0,1$, we assume that the true survival times are observations on a non-negative random variable X_i^0, with absolutely continuous cumulative distribution function F_i. The hazard functions in each sample are defined to be $\lambda_i(t) = \frac{dF_i(t)}{1-F_i(t)}$; we assume that $\lambda_1(t) = \theta\,\lambda_0(t)$ (which implies that $1-F_1 = (1-F_0)^\theta$). Our objective is to investigate several estimators of θ, the relative risk.

In general, the possible follow-up time is limited by a fixed end point to the study, while patients enter a trial at possibly different times. The potential follow-up may thus vary from patient to patient, and may be modeled as non-negative random variables T_i (possibly depending on the sample) with cumulative distribution functions C_i. The T_i are assumed to be independent of the X_i^0 and what is observed is a survival time $X_i = \min\{X_i^0, T_i\}$, as well as whether the observation is uncensored $(X_i = X_i^0)$ or censored $(X_i = T_i)$. (This random censorship model is, in fact, more restrictive than necessary, and is used here primarily for ease of exposition.)

The distribution function of the survival times X_i is given by $H_i = 1-(1-C_i)(1-F_i)$, and the subdistribution function of the observed deaths from each sample will be denoted by G_i, where $dG_i(t) = (1-H_i(t))\,\lambda_i(t)dt$. The number of patients in each sample will be denoted by n_i, with $n = n_0 + n_1$ and $\alpha_i = \lim_{n\to\infty} \frac{n_i}{n}$. Subscript n will be used to denote empirical versions of cumulative distribution functions, and we will also define $G_n = \frac{n_0}{n} G_{0n} + \frac{n_1}{n} G_{1n}$, $H_n = \frac{n_0}{n} H_{0n} + \frac{n_1}{n} H_{1n}$, $G = \alpha_0 G_0 + \alpha_1 G_1$, and $H = \alpha_0 H_0 + \alpha_1 H_1$.

For the combined sample of n patients let the ordered survival times, without regard to censoring, be denoted by

$$\underset{\sim}{X} = (X_1, \ldots, X_n) \quad .$$

Define a corresponding vector of sample indicators

$$\underset{\sim}{Z} = (Z_1,\ldots,Z_n) \quad ,$$

where
$$\begin{array}{ll} Z_j = 0 & X_j \text{ from sample } 0 \\ Z_j = 1 & X_j \text{ from sample } 1 \quad , \end{array}$$

and a vector of censoring indicators

$$\underset{\sim}{\delta} = (\delta_1,\ldots,\delta_n) \quad ,$$

where
$$\begin{array}{ll} \delta_j = 0 & X_j \text{ censored} \\ \delta_j = 1 & X_j \text{ uncensored} \quad . \end{array}$$

Further, define the number at risk in the respective samples at X_j by

$$n_{0j} = \sum_{k=j}^{n} Z_k$$

and
$$n_{1j} = \sum_{k=j}^{n} (1 - Z_k) \quad .$$

3. <u>Likelihood Approaches</u>

3.1 Maximum Likelihood

With an exponential model, we have

$$F_0(t) = 1 - e^{-\lambda_0 t} \quad \text{and} \quad F_1(t) = 1 - e^{-\lambda_1 t} \quad .$$

The likelihood is in this case well known (Bartholomew (1957)) to be maximized by the ratio of occurrences to exposure time:

(1) $$\hat{\lambda}_0 = \sum_{j=1}^{n} \delta_j (1-Z_j) / \sum_{j=1}^{n} X_j (1-Z_j) = \int_0^\infty dG_{0n}(t) / \int_0^\infty t dH_{0n}(t)$$

and $$\hat{\lambda}_1 = \sum_{j=1}^{n} \delta_j Z_j / \sum_{j=1}^{n} X_j Z_j = \int_0^\infty dG_{1n}(t) / \int_0^\infty t dH_{1n}(t) \quad .$$

Thus, the maximum likelihood estimator of θ is $\hat{\lambda}_1/\hat{\lambda}_0$. From (1) it can be seen that, in general,

$$\hat{\lambda}_i \xrightarrow{P} \int_0^\infty dG_i(t) \Big/ \int_0^\infty t dH_i(t)$$

$$= \int_0^\infty \lambda_i(t)\,(1-H_i(t))dt \Big/ \int_0^\infty (1-H_i(t))dt \quad ,$$

so that $\hat{\lambda}_1/\hat{\lambda}_0 \xrightarrow{P} \theta$ under the exponential model, but not necessarily otherwise.

Asymptotic normality of $\hat{\lambda}_1/\hat{\lambda}_0$ can be shown in general using the representation (1) and follows from likelihood theory under the exponential model (see Gardiner (1982) for details.) In the latter case, the large sample variance of $\sqrt{n}\,(\hat{\lambda}_1/\hat{\lambda}_0 - \theta)$ is

(2) $$\frac{\theta^2}{\alpha_0\alpha_1}\left[\frac{\int_0^\infty [\alpha_0(1-H_0(t)) + \alpha_1(1-H_1(t))\,\theta\,]\,\lambda_0(t)\,dt}{\int_0^\infty (1-H_0(t))\,\lambda_0(t)dt \cdot \int_0^\infty (1-H_1(t))\,\theta\,\lambda_0(t)dt}\right] \quad .$$

3.2 Maximum Partial Likelihood

Cox (1972) presented a statistical procedure for inference from censored survival data which depended on the model of proportional hazards, and used a likelihood which did not depend on the form of $\lambda_0(t)$. This is not a likelihood in the standard sense but was later shown by Cox (1975) to be a partial likelihood.

In the two sample case, and using $\ln\theta = \beta$, the Cox partial likelihood is

$$L(\beta) = \prod_{j=1}^{n}[\exp(\beta Z_j) / (n_{0j} + n_{1j} \exp(\beta))]^{\delta_j} ,$$

and $\ln(L(\beta)) =$

(3)
$$\ell(\beta) = \sum_{j=1}^{n} \delta_j[\beta Z_j - \ln(n_{0j} + n_{1j} \exp(\beta))] ,$$

The proposed estimator of θ is then $e^{\hat{\beta}}$, where $\hat{\beta}$ is the solution to

$$0 = \ell'(\beta) = \sum_{j=1}^{n} \delta_j \left[Z_j - \frac{n_{1j} \exp(\beta)}{n_{0j} + n_{1j} \exp(\beta)} \right]$$

(4)
$$= n_1 \int_0^{\infty} dG_{1n}(t) - n \int_0^{\infty} \frac{n_1(1-H_{1n}(t^-)) \exp(\beta)\, dG_n(t)}{n_0(1-H_{0n}(t^-)) + n_1(1-H_{1n}(t^-)) \exp(\beta)} .$$

McRae and Thomas (1972) show there is a solution to (4) corresponding to a maximum of (3) with arbitrarily high probability as $n \to \infty$.

The estimated variance is $-[e^{\hat{\beta}}]^2/\ell''(\hat{\beta})$, where

$$\ell''(\beta) = - \sum_{j=1}^{n} \frac{\delta_j n_{0j} n_{1j} \exp(\beta)}{[n_{0j} + n_{1j} \exp(\beta)]^2}$$

(5)
$$= - \frac{n_1 n_2}{n} \int_0^{\infty} \frac{(1-H_{0n}(t^-))(1-H_{1n}(t^-)) \exp(\beta)\, dG_n(t)}{\left[\frac{n_0}{n} (1-H_{0n}(t^-)) + \frac{n_1}{n} [1-H_{1n}(t^-)) \exp(\beta)\right]^2} ,$$

Thus, $\sqrt{n}\,(e^{\hat{\beta}}-\theta)$ is taken to be asymptotically normal, with mean 0 and variance

(6)
$$\frac{\theta}{\alpha_0\alpha_1} \left[\int_0^{\infty} \frac{[1-H_0(t)][1-H_1(t)]\, \theta\, \lambda_0(t)dt}{\alpha_0(1-H_0(t)) + \alpha_1(1-H_1(t))\, \theta} \right]^{-1} ,$$

the denominator arising from the limit in probability of $-\frac{\ell''(\hat{\beta})}{n}$.

The log-likelihood $\ell(\beta)$ is composed of random variables which are neither independent nor identically distributed, so standard likelihood theory fails to justify the large sample moments and distribution given above. However, an approach is possible following the outline of the proofs for standard likelihoods, as suggested in Cox (1975). There are several articles covering the large sample theory for random covariates (Tsiatis, 1981, Andersen and Gill, 1981, for example), but none which explicitly cover the two-sample case arguing directly from the partial likelihood using classical methods. We sketch such a proof below.

Expanding $\ell'(\beta)$ around the true value β_0, we have

$$\ell'(\hat{\beta}) = 0 = \ell'(\beta_0) + (\hat{\beta} - \beta_0)\, \ell''(\beta^*) \quad ,$$

where β^* is in $(\beta_0, \hat{\beta})$. Thus

$$\sqrt{n}\,(\hat{\beta} - \beta_0) = \frac{-n^{-\frac{1}{2}}\, \ell'(\beta_0)}{\ell''(\beta^*)/n} \quad .$$

The result follows if $n^{-\frac{1}{2}}\, \ell'(\beta_0)$ is asymptotically normal with mean 0 and variance estimated by $-\frac{1}{n}\,\ell''(\beta_0)$, if $\hat{\beta} \xrightarrow{P} \beta_0$, and if $\frac{1}{n}\,(\ell''(\beta^*) - \ell''(\beta_0)) \xrightarrow{P} 0$.

From (4) we can see that $\ell'(\beta_0)$ is just the logrank test, with the addition of the constant $\exp(\beta_0)$ multiplying n_{2j}, and thus trivial extensions of existing proofs show that

$$\frac{n^{-\frac{1}{2}}\, \ell'(\beta_0)}{[\ell''(\beta_0)/n]^{\frac{1}{2}}} \xrightarrow{D} N(0,1) \quad ,$$

under proportional hazards and random censorship (Crowley and Thomas, 1975) or more general censoring (Aalen, 1978). Further, it is clear from (5) that $\frac{1}{n}\,[\ell''(\beta^*) - \ell''(\beta_0)]$ will converge in probability to 0 if $\hat{\beta} \xrightarrow{P} \beta_0$.

To show the consistency of $\hat{\beta}$, note that $\ell''(\beta) < 0$, so that $\ell'(\beta)$ is decreasing, and that $\frac{\ell'(\beta)}{n}$ can be seen from (4) to converge in probability to

$$\alpha_1 \int_0^\infty dG_1(t) - \int \frac{\alpha_1(1-H_1(t)) \exp(\beta)\, dG(t)}{\alpha_0(1-H_0(t)) + \alpha_1(1-H_1(t)) \exp(\beta)} \quad . \tag{7}$$

If the true parameter is β_0, and setting $\beta = \beta_0 + \Delta$, (7) decomes

$$\int_0^\infty \alpha_1(1-H_1(t)) \exp(\beta_0)\, \lambda_0(t)dt - \int \frac{\alpha_1(1-H_1(t)) \exp(\beta_0 + \Delta)(1-F_0(t))^{\exp(\Delta)}\, dG(t)}{\alpha_0(1-H_0(t)) + \alpha_1(1-H_1(t)) \exp(\beta_0 + \Delta)} \quad . \tag{8}$$

Writing $dG(t) = [\alpha_0(1-H_0(t) + \alpha_1(1-H_1(t)) \exp(\beta_0)]\, \lambda_0(t)dt$, (8) simplifies after a little algebra to

$$= \alpha_0\alpha_1 \int_0^\infty \frac{(1-H_0(t))(1-H_1(t)) \exp(\beta_0)(1-\exp(\Delta))\lambda_0(t)dt}{\alpha_0(1-H_0(t)) + \alpha_1(1-H_1(t)) \exp(\beta_0 + \Delta)} \quad . \tag{9}$$

The expression (9) is 0 when $\Delta = 0$, and negative (positive) when Δ is positive (negative); there is thus with high probability a single root of $\ell'(\beta) = 0$ in an arbitrary neighborhood of β_0, so that $\hat{\beta} \overset{P}{\rightarrow} \beta_0$.

This result can be generalized to vector-valued covariates, not necessarily i.i.d., with a finite number of possible outcomes. The approach requires establishing that

$$\lim_{n \to \infty} P\{(\ell(\beta_0) - \sup_{\beta \in N_\Delta(\beta_0)} \ell(\beta)) > 0\} = 1 \quad ,$$

for $N_\Delta(\beta_0) = \{\beta : |\beta - \beta_0| = \Delta\}$. This rests on careful consideration of the terms $U_j(\beta)$ of the log-partial likelihood, centered by their conditional expectations under $\beta = \beta_0$ given the information on the times of all previous censored and uncensored observations and the associated covariates, and on the fact that an uncensored observation occurs at X_j. Details are given in Liu and Crowley (1978).

4. Ad Hoc Methods

4.1 Standardized Mortality Ratio

Peto, et al. (1977) provide an excellent review of the analysis of clinical trials and point out the need for simple, closed-form test statistics and estimators. They suggest that by analogy with the standardized mortality ratio, an estimator of relative risk is provided by the ratio of observed deaths in sample 1 to that expected under the null hypothesis. This is

$$O/E = \frac{\sum_{j=1}^{n} \delta_j Z_j}{\sum_{j=1}^{n} \delta_j \frac{n_{1j}}{n-j+1}} ;$$

n-j+1 being the total number at risk at X_j and $n_{1j}/n-j+1$ thus the estimated number of deaths from sample 1. This can be rewritten as

$$\frac{\int_0^\infty dG_{1n}(t)}{\int_0^\infty \frac{1-H_{1n}(t^-)}{1-H_n(t^-)} \, dG_n(t)} ,$$

from which it can be seen that

$$O/E \xrightarrow{P} \frac{\int_0^\infty dG_1(t)}{\int_0^\infty \frac{(1-H_1(t))\, dG(t)}{(1-H(t))}} .$$

Under proportional hazards this is

$$\text{(10)} \qquad \frac{\theta \int_0^\infty (1-H_1(t))\, \lambda_0(t)dt}{\int_0^\infty \frac{(1-H_1(t))\, [\alpha_0(1-H_0(t)) + \alpha_1(1-H_1(t))\,\theta\,]\, \lambda_0(t)dt}{\alpha_0(1-H_0(t)) + \alpha_1(1-H_1(t))}} .$$

The estimator O/E is thus biased towards 1. For example, with the exponential model and no censoring, (10) is equal to 1.24 when $\theta = 1.5$, and 1.44 when $\theta = 2$. However, Bernstein, Anderson, and Pike (1981) present some Monte Carlo results which indicate that O/E behaves fairly well for moderate sample sizes and moderate departures from the null hypothesis.

4.2 The Mantel-Haenszel Estimator

An analogy with the analysis of case-control studies could also be drawn, as was done for the logrank (or Mantel-Haenszel) test by Mantel (1966). This would suggest that the Mantel-Haenszel (1959) estimator of the log odds ratio (relative risk for rare diseases) from sets of 2×2 tables be used as an estimator of relative risk for survival studies as well. This is given by

$$\hat{\theta}_R = \frac{\sum_{j=1}^{n} \frac{\delta_j Z_j (n_{0j} - (1-Z_j))}{n-j+1}}{\sum_{j=1}^{n} \frac{\delta_j (1-Z_j)(n_{1j} - Z_j)}{n-j+1}}$$

$$= \frac{\displaystyle\int_0^\infty \frac{(1-H_{0n}(t))\ dG_{1n}(t)}{(1-H_n(t^-))}}{\displaystyle\int_0^\infty \frac{(1-H_{1n}(t))\ dG_{0n}(t)}{(1-H_n(t^-))}} .$$

In establishing the large sample properties of $\hat{\theta}_R$ it is convenient to use the results on counting processes, outlined, for example, by Andersen, Borgan, Gill and Keiding (1981). Thus, $n_i\ dG_{in}(t)$ are counting processes with compensators $\int_0^t \lambda_i(s)\ n_i(1-H_{in}(s^-))ds$, and

$$M_i(t) = n_i\ G_{in}(t) - \int_0^t \lambda_i(s)\ n_i(1-H_{in}(s^-))ds \quad ,$$

are orthogonal square-integrable martingales. Also, as $\sqrt{n}\ (1-H_{1n}(t^-))/ n_0(1-H_n(t^-))$ and $\sqrt{n}\ (1-H_{0n}(t^-))/n_1(1-H_n(t^-))$ are left-continuous and therefore predictable processes, and are bounded (interpreting 0/0 as 0), we have that

$$\sqrt{n} \int_0^\infty \frac{(1-H_{1n}(t^-))}{n_0(1-H_n(t^-))}\ dM_0(t)$$

and

$$\sqrt{n} \int_0^\infty \frac{(1-H_{0n}(t^-))}{n_1(1-H_n(t^-))}\ dM_1(t) \quad ,$$

are orthogonal square-integrable martingales. Further, the conditions for asymptotic normality given by Andersen, Borgan, Gill and Keiding (1981) are satisfied, and we have that

$$\sqrt{n} \left[\int_0^\infty \frac{(1-H_{1n}(t^-))}{(1-H_n(t^-))}\ dG_{0n}(t) - \int_0^\infty \frac{(1-H_{1n}(t^-))}{(1-H_n(t^-))}\ \lambda_0(t)(1-H_{0n}(t^-))dt \right] ,$$

and

$$\sqrt{n}\left[\int_0^\infty \frac{(1-H_{0n}(t^-))}{(1-H_n(t^-))}\,dG_{1n}(t) - \int_0^\infty \frac{(1-H_{0n}(t^-))}{(1-H_n(t^-))}\,\lambda_1(t)(1-H_{1n}(t^-))dt\right] ,$$

are in the limit, mean 0, independent normal random variables with variances

$$\sigma_0^2 = \frac{1}{\alpha_0}\int_0^\infty \frac{(1-H_0(t))(1-H_1(t))^2}{(1-H(t))^2}\,\lambda_0(t)dt$$

$$\sigma_1^2 = \frac{1}{\alpha_1}\int_0^\infty \frac{(1-H_0(t))^2(1-H_1(t))}{(1-H(t))^2}\,\lambda_1(t)dt \quad .$$

Thus, we have a result of the form

$$\sqrt{n}\,(A_{0n} - B_{0n},\, A_{1n} - B_{1n}) \overset{D}{\to} N\left(0, \begin{pmatrix} \sigma_0^2 & 0 \\ 0 & \sigma_1^2 \end{pmatrix}\right) ,$$

and our estimator $\hat{\theta}_R$ can be seen to be $\frac{A_{1n}}{A_{0n}}$ (except for the presence of t^- instead of t in the respective numerators, which will not matter in the limit because of the assumed continuity of $F_i(t)$). Asymptotic normality of $\hat{\theta}_R$ follows from repeated application of the δ-method, expanding A_{0n} and B_{0n} around their common limit

$$\mu_0 = \int_0^\infty \frac{(1-H_1(t))}{(1-H(t))}\,dG_0(t) \quad ,$$

and A_{1n} and B_{1n} around

$$\mu_1 = \int_0^\infty \frac{(1-H_0(t))}{(1-H(t))}\,dG_1(t) \quad .$$

This gives

$$\sqrt{n}\left(\frac{A_{1n}}{A_{0n}} - \frac{B_{1n}}{B_{0n}}\right) \xrightarrow{D} N\left(0, \left(\frac{\mu_1}{\mu_0}\right)^2 \cdot \left(\frac{\sigma_0^2}{\mu_0^2} + \frac{\sigma_1^2}{\mu_1^2}\right)\right).$$

For proportional hazards, $\frac{B_{1n}}{B_{0n}} = \frac{\mu_1}{\mu_0} = \theta$, and we have established that $\hat{\theta}_R \xrightarrow{P} \theta$, and that $\sqrt{n}\,(\hat{\theta}_R - \theta)$ is asymptotically normal, with variance which reduces to

$$\text{(11)}\qquad \frac{\theta^2}{\alpha_0\alpha_1}\left\{\frac{\displaystyle\int_0^\infty \frac{(1-H_0(t))(1-H_1(t))}{(1-H(t))^2}\,[\alpha_0(1-H_0(t)) + \alpha_1(1-H_1(t))\theta]\;\lambda_0(t)dt}{\theta\left[\displaystyle\int_0^\infty \frac{(1-H_0(t))(1-H_1(t))}{1-H(t)}\,\lambda_0(t)dt\right]^2}\right\}$$

The estimator $\hat{\theta}_R$ has been generalized to statistics of the form

$$\text{(12)}\qquad \frac{\displaystyle\int_0^\infty J(H_{0n}(t),\, H_{1n}(t))\,\frac{dG_{1n}(t)}{1-H_{1n}(t^-)}}{\displaystyle\int_0^\infty J(H_{0n}(t),\, H_{1n}(t))\,\frac{dG_{0n}(t)}{1-H_{0n}(t^-)}},$$

which will also be consistent for θ under the proportional hazards model and will be asymptotically normal for a certain class of functions J. The choice of $J = \frac{(1-H_{0n}(t^-))(1-H_{1n}(t^-))}{1-H_n(t^-)}$ corresponds to $\hat{\theta}_R$. This is discussed from the point of view of the resulting test statistics by Andersen (1981), and from the point of view of estimation by Begun and Reid (1981). That some restrictions on J are necessary can be seen from the choice of $J\equiv 1$, resulting in the ratio of cumulative hazards

$$\frac{\displaystyle\int_0^\infty \frac{dG_{1n}(t)}{1-H_{1n}(t^-)}}{\displaystyle\int_0^\infty \frac{dG_{0n}(t)}{1-H_{0n}(t^-)}} = \frac{\displaystyle\sum_{j=1}^{n} \frac{\delta_j z_j}{n_{ij}}}{\displaystyle\sum_{j=1}^{n} \frac{\delta_j (1-z_j)}{n_{0j}}} = \frac{\hat{\Lambda}_1(\infty)}{\hat{\Lambda}_0(\infty)} ,$$

where $\hat{\Lambda}_i(t)$ is sometimes referred to as Nelson's estimator (Nelson, 1969) of the cumulative or integrated hazard $\int_0^t \lambda_i(s)ds$. The ratio may well provide a useful estimate as a function of time, for

$$\frac{\hat{\Lambda}_1(t)}{\hat{\Lambda}_0(t)} = \frac{\hat{\Lambda}_1(t)/t}{\hat{\Lambda}_0(t)/t} ,$$

has the interpretation of the ratio of time-averaged hazards, but choice of t will be important in its use as an estimator, as the asymptotic variance of $\hat{\Lambda}(t)$ can increase without bound (cf Breslow and Crowley, 1974). A similar point was made by Kalbfleisch and Prentice (1981), who study estimates of the average hazard ration, defined to by

$$\int_0^t \lambda_1(s)/\lambda_0(s)\, dW(s) ,$$

for suitably chosen weight functions W.

5. Some Numerical Comparisons

As a large sample measure of efficiency we can compare asymptotic variances for those estimators which are consistent. With the exponential model this includes the maximum likelihood estimator $\hat{\lambda}_1/\hat{\lambda}_0$, the maximum partial likelihood estimator $e^{\hat{\beta}}$, and the Mantel-Haenszel estimator $\hat{\theta}_R$. Since it is maximum likelihood, we have that A.Var. $\sqrt{n}\,(\hat{\lambda}_1/\hat{\lambda}_0 - \theta) \leq$ A.Var. $\sqrt{n}\,(e^{\hat{\beta}} - \theta)$, where A.Var. stands for asymptotic variance. For the general proportional hazards model we can see from (6) and (11) and the Cauchy-Schwarz inequality that A.Var.

$\sqrt{n}\,(e^{\hat{\beta}} - \theta) \leq$ A.Var. $\sqrt{n}\,(\hat{\theta}_R - \theta)$, with equality with $\theta=1$. With equal exponential survival distributions and equal censoring, comparison of (2) with (6) and (11) reveals that all three estimators have the same asymptotic variance.

Table 1 gives large sample variances for the three estimators under an exponential model with $\lambda_0 = 1$, covering the cases $\theta = 1, 1.5$, and 2.0 for four different censoring patterns. Conditions 1 and 2 correspond to a cohort entering the study at time 0, with staggered entry of other subjects from 0 to 1 and analysis at time 1; case 1 having an equal size cohort in each sample and case 2, unequal. The third censoring condition is equal, type I censoring at time 1, and the fourth is no censoring. Also given in Table 1 are comparisons of $e^{\hat{\beta}}$ and $\hat{\theta}_R$ for Weibull distributions under the same censoring conditions. The most remarkable feature of the table is the high relative efficiency of the simple, closed-form Mantel-Haenszel estimator. This small loss can be regained entirely in a two-step procedure suggested by Begun and Reid (1981), using the statistic (12). They show that the optimal J depends on θ, but that use of this J with θ replaced by $\hat{\theta}_R$ (or any consistent estimator) provides full efficiency relative to the maximum partial likelihood estimator.

Further numerical comparisons covering the case of grouped survival times are given in Crowley (1975).

TABLE 1. Large Sample Variances

I. $1-F_0(t) = e^{-t}$ $1-F_1(t) = e^{-\theta t}$

1) $C_i(t) = t/1.25$, $t\varepsilon[0,1];1,\ t > 1$

θ	$\hat{\lambda}_1/\hat{\lambda}_0$	$e^{\hat{\beta}}$	$\hat{\theta}_R$
1	9.51	9.51	9,51
1.5	19.01	19.08	19.08
2	31.77	32.20	32.25

2) $C_0(t) = t/1.25$ $C_1(t) = t/1.5$, $t\varepsilon[0,1];1,\ t > 1$

θ	$\hat{\lambda}_1/\hat{\lambda}_0$	$e^{\hat{\beta}}$	$\hat{\theta}_R$
1	9.14	9.16	9.16
1.5	18.45	18.46	18.46
2	31.02	31.21	31.24

3) $C_0(t) = C_1(t) = 0$, $t\varepsilon[0,1];1,\ t > 1$

θ	$\hat{\lambda}_1/\hat{\lambda}_2$	$e^{\hat{\beta}}$	$\hat{\theta}_R$
1	6.33	6.33	6.33
1.5	12.91	12.97	12.98
2	21.91	22.31	22.36

4) No censoring

θ	$\hat{\lambda}_1/\hat{\lambda}_2$	$e^{\hat{\beta}}$	$\hat{\theta}_R$
1	4	4	4
1.5	9	9.35	9.36
2	16	17.75	17.87

Table 1 (continued)

II. $1-F_0(t) = e^{-t^2}$ $1-F_1(t) = e^{-\theta t^2}$

1) $C_i(t) = t/1.25$, $t \varepsilon [0,1]; 1,\ t > 1$

θ	$e^{\hat{\beta}}$	$\hat{\theta}_R$
1	12.16	12.16
1.5	24.37	24.37
2	41.07	41.14

2) $C_0(t) = t/1.25$, $C_1(t) = t/1.5$, $t \varepsilon [0,1]; 1,\ t > 1$

θ	$e^{\hat{\beta}}$	$\hat{\theta}_R$
1	11.38	11.38
1.5	22.99	22.99
2	38.90	38.92

Results for cases 3) and 4) are the same as for corresponding entries in Part I.

REFERENCES

Aalen, O.O. (1978). Nonparametric inference for a family of counting processes. Annals of Statistics 6, 374-391.

Andersen, P.K. (1981). Comparing survivals via hazard ratio estimates. Research Report 81/7, Statistical Research Unit, Danish Medical and Social Science Research Councils. To appear in the Scandinavian Journal of Statistics, 1983.

Anderson, P.K. and Gill, R. (1981). Cox's regression model for counting processes: a large sample study. Research Report 81/6, Statistical Research Unit, Danish Medical and Social Science Research Councils. To appear in the Annals of Statistics, 1982.

Anderson, P.K., Borgan, Ø., Gill, R.D. and Keiding, N. (1981). Linear nonparametric tests for comparison of counting processes, with applications to censored survival data. Research Report 81/4, Statistical Research Unit, Danish Medical and Social Science Research Councils. To appear in the International Statistical Review, 1982.

Bartholomew, D.J. (1957). A problem in life testing. Journal of the American Statistical Association 52, 350-355.

Begun, J.M. and Reid, N. (1981). Estimating the relative risk with censored data. Technical Report No. 81/16, The Institute of Applied Mathematics and Statistics, the University of British Columbia. To appear in the Journal of the American Statistical Association, 1983.

Bernstein, L., Anderson, J., and Poke, M.C. (1981). Estimation of the proportional hazard in two-treatment-group clinical trials. Biometrics 37, 513-519.

Breslow, N. and Crowley, J. (1974). A large sample study of the life table and product limit estimates under random censorship. Annals of Statistics 2, 437-453.

Cox, D.R. (1972). Regression models and life tables (with discussion). Journal of the Royal Statistical Society B 34, 187-220.

Cox, D.R. (1975). Partial likelihood. Biometrika 62, 269-276.

Crowley, J. and Thomas, D.R. (1975). Large sample theory for the log rank test. Technical Report No. 415, University of Wisconsin-Madison, Department of Statistics.

Crowley, J. (1975). Estimation of relative risk in survival studies. Technical Report No. 423, University of Wisconsin-Madison, Department of Statistics.

Gardiner, J. (1982). The asymptotic distribution of mortality rates in competing risks analysis. Scandinavian Journal of Statistics 9, 31-36.

Kalbfleisch, J.D. and Prentice, R.L. (1981). Estimation of the average hazard ratio. Biometrika 68, 105-112.

Liu, P.Y. and Crowley, J. (1978). Large sample theory of the MLE based on Cox's regression and life model for survival data. Technical Report No. 1, Wisconsin Clinical Cancer Center - Biostatistics.

Mantel, N., and Haenszel, W. (1959). Statistical aspects of the analysis of data from retrospective studies of disease. Journal of the National Cancer Institute 22, 719-748.

Mantel, N. (1966). Evaluation of survival data and two new rank order statistics arising in its consideration. Cancer Chemotherapy Reports 50, 163-170.

McRae, K.B. and Thomas, D.R. (1972). Inference for pairs of distributions with proportional failure rate functions. Paper presented at the Joint Statistical Meetings in Montreal.

Nelson, W. (1969). Hazard plotting for incomplete failure data. Journal of Quality Technology 1, 27-52.

Peto, R., Pike, M.C., Armitage, P., Breslow, N.E., Cox, D.R., Howard, S.V., Mantel, N., McPherson, K., Peto, J., and Smith, P.G. (1977). Design and analysis of randomized clinical trials requiring prolonged observation of each patient. II. Analysis and examples. British Journal of Cancer 35, 1-39.

Tsiatis, A.A. (1981). A large sample study of Cox's regression model. Annals of Statistics 9, 93-108.

ON THE PERFORMANCE OF ESTIMATES IN PROPORTIONAL HAZARD AND LOG-LINEAR MODELS

Kjell A. Doksum

Department of Statistics, University of California, Berkeley

1. Introduction

Let $T_1, T_2, \ldots, T_n$ be independent survival times with T_i having distribution function (d.f.) F_i, density f_i and hazard rate $\lambda_i(t) = f_i(t)/[1-F_i(t)]$.

One model often used in the analysis of survival experiments is the proportional hazard model where

$$\lambda_i(t) = \Delta_i \lambda(t) \ , \quad t \geq 0 \tag{1}$$

for some constant $\Delta_i > 0$. Here $\lambda(t) = f(t)/[1-F(t)]$ for d.f. F with density f. In a different context, this model was considered by Lehmann (1953) and Savage (1956) in the equivalent form $F_i(t) = 1-[1-F(t)]^{\Delta_i}$, some d.f. F. It was used by Cox (1972) in situations where the distribution of T_i depends on p covariates $x_{i1}, \ldots, x_{ip}$. Cox modeled this dependence by assuming

$$\lambda_i(t) = \Delta_i \lambda(t) \ , \quad \Delta_i = \exp\left(\sum_{j=1}^{p} x_{ij} \beta_j \right) \ , \tag{2}$$

where $\underset{\sim}{\beta} = (\beta_1, \ldots, \beta_p)^T$ is a vector of regression coefficients.

Another model often used with survival distributions is the scale model where

$$F_i(t) = G(t/\tau_i) \ , \quad \text{some } \tau_i > 0 \ , \quad \text{some d.f. } G \ . \tag{3}$$

When F_i depends on covariates, one way to model this dependence is by writing

(4) $$\log T_i = \sum_{j=1}^{p} x_{ij}\theta_j + e_i \quad .$$

Here $\underset{\sim}{x} = (x_{ij})$ are the same covariates as before and $\underset{\sim}{\theta} = (\theta_1,\ldots,\theta_p)^T$ is a vector of regression coefficients. Note that (4) is a special case of (3) with $\tau_i = \exp(\sum_{j=1}^{p} x_{ij}\theta_j)$ and $G(t) = H(\log t)$ where H is the d.f. of e_i.

In certain studies, there will be cenoring variables $C_1,\ldots,C_n$, and one observes $T_i' = \min(T_i,C_i)$, and $\delta_i = I[T_i \leq C_i]$, rather than T_i, where I is the indicator function.

Cox (1972) has introduced partial likelihood estimates for the model (2); and Miller (1976), Buckley and James (1979), and Koul, Susarla and Van Ryzin (1981) have considered least squares type estimators for the model (4).

In the next sections, we will first show that the models (1) and (3) coincide only for the Weibull model and then make asymptotic comparisons between the Cox estimates, least squares type estimates and rank estimates. In the Weibull model, the rank estimates are asymptotically optimal. Efficiency results are obtained in very special cases.

2. The Equivalence of the Proportional Hazard and Log-linear Models

The result that the proportional hazard and log-linear models coincide only when T_i has a Weibull distribution has appeared in Doksum (1975), Kalbfleisch and Prentice (1980, p.34) and Louis (1981). Only the second reference gives a proof and in this proof the covariates $\underset{\sim}{x}$ are allowed to vary and in fact are allowed to be functions of the regression coefficients.

We give a different proof which requires that (i) the proportional hazard model (1) and scale model (3) coincide when τ_i and Δ_i are unity, and (ii) that they coincide for at least one value of τ_i different from unity. We also need the regularity condition: (iii) For some $a > -1$, $\lim_{t\to 0^+} [\lambda(t)/t^a]$ exists and is positive.

The proof proceeds as follows: From (i) we conclude that G in (3) equals the F in $\lambda(t) = f(t)/[1-F(t)]$ of model (1). Now (3) and (ii) implies that $\lambda_i(t) = \lambda(t/\tau_i)/\tau_i$ for some $\tau_i \neq 1$. When this is combined with (1), we obtain

$$(5) \qquad \lambda(t/\tau) = \tau\Delta\lambda(t) \ , \quad \text{all } t \ , \quad \text{some } \tau \neq 1 \ ,$$

where we dropped the subscripts on τ_i and Δ_i. We will show that (5) implies that $\lambda(t)$ must be the failure rate of a Weibull distribution, i.e., that $\lambda(t)$ is of the form $\lambda(t) = ct^{\gamma-1}$, some $\gamma > 0$.

First suppose that $0 < \lambda(0) < \infty$, then $\lambda(0) = \Delta\tau\lambda(0)$ implies $\Delta\tau = 1$, i.e., $\lambda(t/\tau) = \lambda(t)$ for all $t \geq 0$. Now when $\tau \neq 1$ and (iii) holds, this implies $\lambda(t) =$ constant, i.e., the model is exponential.

Next suppose that $\lambda(0) = 0$. Let $h(t) = \lambda(t)/t^a$ where a is given in (iii). Using (5), we find $h(t/\tau) = \Delta\tau^{a+1}h(t)$. Now since $0 < h(0^+) < \infty$, $h(0^+) = \Delta\tau^{a+1}h(0^+)$ implies $\Delta\tau^{a+1} = 1$, thus $h(t/\tau) = h(t)$ for all $t > 0$. Since $\tau \neq 1$, this implies $h(t) =$ constant, i.e., $\lambda(t) = ct^a$, and the model is Weibull.

Equations that include equation (5) as a special case can be found in Kuczma (1968, p.47) and Nabeya (1974), but the present solution is not given there.

In the Weibull model we use the notation where T_i had d.f.

$$(6) \qquad F_i(t) = 1 - \exp\{-(t/\tau_i)^\gamma\} \ .$$

Here $\log \tau_i = \sum_{j=1}^{p} x_{ij}\theta_j$ as in (4). In the Cox model (2), the model (6) corresponds to $\lambda(t) = t^{\gamma-1}$, $\Delta_i = \tau_i^{-\gamma}$. Thus $\log \tau_i = -\gamma^{-1} \sum_{j=1}^{p} x_{ij}\beta_j$, and the correspondence between $\underset{\sim}{\theta}$ and $\underset{\sim}{\beta}$ in the Weibull model is $\underset{\sim}{\theta} = -\gamma^{-1}\underset{\sim}{\beta}$.

3. The Estimates

3.1 Least Squares (L.S.) Type Estimates

We consider only the uncensored case. The asymptotic variance of L.S. type estimates has been obtained for certain types of censoring by Miller (1976) and Koul, Susarla and Van Ryzin (1981). We consider the model (4) with $\underset{\sim}{x}$ of full rank and $e_1,\ldots,e_n$ i.i.d. The variance of the L.S. estimate $\hat{\underset{\sim}{\theta}} = (\hat{\theta}_1,\ldots,\hat{\theta}_p)$ is then $\sigma^2(x^T x)^{-1}$, where $\sigma^2 = \text{Var}(e_i)$. If we specialize to the Weibull model (6) we find that e_i has d.f. H given by

$$H(t) = 1 - \exp[-\exp(\gamma t)] \quad ,$$

and variance $\text{Var}(e_i) = \text{Var}(\log T_i) = \pi^2/6\gamma^2$.

Note that $E(e_i)$ is not equal to zero; in fact $E(e_i) = -E/\gamma$ where E = Euler's constant $\cong .5772$. It follows that the L.S. type estimates are not necessarily consistent for the Weibull model. Thus, if $p = 2$, and $\log T_i = \theta_1 + \theta_2 x_i + e_i$, then $\hat{\theta}_1$ converges in probability to $\theta_1 - E/\gamma$ while $\hat{\theta}_2$ is consistent. This can be "fixed" by reparametrizing: Set $\log T_i = \theta_1' + \theta_2 x_i + e_i'$, where $e_i' = e_i - E(e_i)$ and $\theta_1' = \theta_1 + E(e_i)$. Note that $E(e_i)$ is unknown if γ is, but we can think of the L.S. estimate as an estimate of the intercept after the errors have been adjusted to have mean zero.

3.2 Cox Estimates

Relevant asymptotic results can be found in papers by Efron (1977), Oakes (1977), Aalen (1978, 1980), Bailey (1979), Tsiatis (1981), and the book by Kalbfleisch and Prentice (1980). In the model (2) with no censoring, let $t_{(1)} < \cdots < t_{(n)}$ be the ordered observed survival times and let $\underset{\sim}{x}_{(i)} = (x_{(i)1},\ldots,x_{(i)p})$ be the covariates corresponding to $t_{(i)}$. Then the Cox estimate $\hat{\underset{\sim}{\beta}} = (\hat{\beta}_1,\ldots,\hat{\beta}_p)$ is the value that maximizes the Cox partial likelihood

$$L_c = \prod_{i=1}^{n} \left\{ \frac{\exp \underset{\sim}{x}_{(i)}\underset{\sim}{\beta}}{\sum_{s=i}^{n} \exp \underset{\sim}{x}_{(s)}\underset{\sim}{\beta}} \right\} \quad .$$

The asymptotic covariance matrix of $(\hat{\beta}_1,\ldots,\hat{\beta}_p)$ is the inverse of the expected value of the observed Cox partial information matrix defined by

$$I^c_{k\ell} = \sum_{i=1}^{n} \left\{ \frac{\sum_{j=i}^{n} x_{(j)k} x_{(j)\ell} \exp\{\underset{\sim}{x}_{(j)} \underset{\sim}{\beta}\}}{\sum_{j=i}^{n} \exp\{\underset{\sim}{x}_{(j)} \underset{\sim}{\beta}\}} - \frac{\left(\sum_{j=i}^{n} x_{(j)k} \exp\{\underset{\sim}{x}_{(j)} \underset{\sim}{\beta}\}\right)\left(\sum_{j=i}^{n} x_{(j)\ell} \exp\{\underset{\sim}{x}_{(j)} \underset{\sim}{\beta}\}\right)}{\left(\sum_{j=i}^{n} \exp\{\underset{\sim}{x}_{(j)} \underset{\sim}{\beta}\}\right)^2} \right\} .$$

Note that the only quantity that is random in this expression is the index (j) in $\underset{\sim}{x}_{(j)}$, $x_{(j)k}$ and $x_{(j)\ell}$.

3.3 Rank Estimates

We consider the log-linear model (4). Properties of estimates based on ranks were developed for the two-sample problem by Hodges and Lehmann (1963), considered for Type II censoring by Doksum (1967), extended to simple linear regression by Adichie (1967), and to multiple linear regression by Jureckova (1971). Let R_i be the rank of T_i among $T_1,\ldots,T_n$. Since $\text{Rank}(T_i) = \text{Rank}(\log T_i)$, the rank approach to log-linear models reduces the estimation problem to the problem of estimating the parameters in a linear model for $Y_i = \log T_i$. The idea in the above references is to use the estimates $\hat{\theta}_1,\ldots,\hat{\theta}_p$ obtained by "inverting" linear rank statistics of the type

$$S_j = \frac{1}{n} \sum_{i=1}^{n} (x_{ij} - x_{\cdot j})\, J_n\left(\frac{R_i}{n+1}\right) ; \quad j=1,\ldots,p$$

where $J_n(\frac{1}{n+1}), \ldots, J_n(\frac{n}{n+1})$ are given scores (constants) and $x_{\cdot j} = \frac{1}{n}\sum_{i=1}^{n} x_{ij}$. When H_0: $\underset{\sim}{\theta} = \underset{\sim}{0}$ holds, the distribution of $\underset{\sim}{S} = (S_1,\ldots,S_p)$ tends to be concentrated near its mean $E_{H_0}(\underset{\sim}{S}) = \underset{\sim}{0}$. When $\underset{\sim}{\theta} \neq \underset{\sim}{0}$, let R_i^{θ} denote the rank of $Y_i - \underset{\sim}{x}_i\underset{\sim}{\theta}$, where $\underset{\sim}{x}_i = (x_{i1},\ldots,x_{ip})$, and let $S_j(\underset{\sim}{Y} - \underset{\sim}{x}\underset{\sim}{\theta}) = \frac{1}{n}\sum_{i=1}^{n} (x_{ij} - x_{\cdot j})\, J_n(\frac{R_i^{\theta}}{n+1})$. When $\underset{\sim}{\theta}$ is the true value of the parameter, the distribution of $S_j(\underset{\sim}{Y} - \underset{\sim}{x}\underset{\sim}{\theta})$ will be concentrated near zero, thus the idea is to use the estimate $\hat{\underset{\sim}{\theta}}$ which "solves" $S_j(\underset{\sim}{Y} - \underset{\sim}{x}\underset{\sim}{\theta}) = 0$, $j=1,\ldots,p$, for $\underset{\sim}{\theta}$. Exact definitions and conditions are in the above references.

Note that since the ranks are invariant under additions of constants, i.e., $\text{Rank}(\log T_i + a) = \text{Rank}(\log T_i)$, this approach can not be used to estimate α in the model $\log T_i = \alpha + \beta x_i + e_i$. Adichie (1967) and Jureckova (1971) introduce rank estimates for α. We do not treat those here.

$\hat{\underset{\sim}{\theta}}$ is consistent and $(\hat{\underset{\sim}{\theta}}-\underset{\sim}{\theta})$ (standardized) is asymptotically normal with mean zero. $\hat{\underset{\sim}{\theta}}-\underset{\sim}{\theta}$ has approximate covariance matrix

$$A^2 B^{-2} (\underset{\sim}{x}^T \underset{\sim}{x})^{-1}$$

where $A^2 = \int_0^1 J^2(u)du - [\int_0^1 J(u)du]^2$, $B = \int_{-\infty}^{\infty} [\frac{d}{dx} J(H(x))]dH(x)$, and J is the limiting score function, $J(u) = \lim_{n\to\infty} J_n(\frac{[nu]+1}{n+1})$, $0 < u < 1$. Here [ ] is the greatest integer function.

Let $\sigma^2(\hat{\underset{\sim}{\theta}};J,G)$ denote the asymptotic variance vector of $[(\hat{\theta}_1 - \theta_1)/b_1, \ldots, (\hat{\theta}_p - \theta_p)/b_p]$ where $b_j = [\sum_{i=1}^{n} (x_{i1} - x_{\cdot j})^2]^{-1/2}$. If the distribution G is known, and thus H is also known, $\sigma^2(\hat{\underset{\sim}{\theta}};J,G)$ is minimized by choosing $J_n(\frac{i}{n+1}) = E[\phi(U^{(i)}, H)]$, where $\phi(u,H) = -h'(H^{-1}(u))/h(H^{-1}(u))$, h is the density of H, and $U^{(i)}$ is the ith uniform order statistic in a sample of size n. Another optimal choice is the simpler form $J_n(\frac{i}{n+1}) = \phi(\frac{i}{n+1},H)$. These results follow from Hájek and Šidák (1967) and the above references.

In particular, if T_i has the Weibull distribution (6), then the optimal J_n is $J_n(\frac{i}{n+1}) = E\{-\log(1-U^{(i)})\} = \sum_{j=n+1-i}^{n} (1/j)$, the exponential or Savage scores. The asymptotically equivalent simpler version is $J_n(\frac{i}{n+1}) = -\log\{1-[i/(n+1)]\}$. Note that these functions do not depend on the shape parameter γ of the Weibull model, thus the exponential scores estimate $\hat{\underset{\sim}{\theta}}_R$ obtained by setting $J_n(\frac{i}{n+1}) = \sum_{j=n+1-i}^{n} (1/j)$ or $J_n(\frac{i}{n+1}) = -\log\{1-[i/(n+1)]\}$ minimizes the asymptotic variance uniformly in γ. This optimality does not hold only in the class of rank estimates, but over the class of all "regular" estimates including least squares and Cox estimates.

The exponential scores estimate has another strong optimality property. It is asymptotically minimax over the class of increasing failure rate average

(IFRA) distributions. More precisely, let $\sigma^2(\hat{\underset{\sim}{\theta}};J) = \sup_G \sigma^2(\hat{\underset{\sim}{\theta}},J,G)$, where the sup is over all G continuous and IFRA. The estimate which minimizes $\sigma^2(\hat{\underset{\sim}{\theta}};J)$ is the exponential scores estimate; moreover for this estimate, the maximum approximate variance (i.e., the maximum of A^2B^{-2}) is attained at the exponential distribution. In fact the approximate covariance matrix $\Sigma(G)$ of $\hat{\underset{\sim}{\theta}}_R$ is such that for the exponential distribution, it reduces to the familiar matrix $(\underset{\sim}{x}^T\underset{\sim}{x})^{-1}$. Thus we can think of $(\underset{\sim}{x}^T\underset{\sim}{x})^{-1}$ as a lower bound for the covariance matrix of $\hat{\underset{\sim}{\theta}}_R$ for IFRA distributions. This result leads immediately to simple bounds on the standard error of $\hat{\underset{\sim}{\theta}}_R$ and confidence intervals for $\underset{\sim}{\theta}$. These results are extensions of Doksum (1967).

Rank estimates for Type II censoring was considered in the two sample case by Doksum (1967). Rank test statistics for censored samples have been considered by Rao-Savage and Sobel (1960), Gastwirth (1965), Gehan (1965), Mantel (1966), Efron (1967), Basu (1968), Doksum (1969), Johnson and Mehrotra (1972), Peto and Peto (1972), Cox (1972), Crowley (1974), Prentice (1978) and Aalen (1978), among others.

The following is an extension of the exponential scores statistic: The survival times are ordered. The first survival time is given score $= \frac{1}{n_1}$, the $(k+1)^{st}$ is given the score of the kth plus the reciprocal of the number n_{k+1} of subjects at risk right before the $(k+1)^{st}$ death. The censoring time C_i is given the score of the largest survival time T to the left of C_i plus one. "One" can be interpreted as the average of the possible scores to the right of (and including) the score of T. If there is no survival time T to the left of C_i, C_i gets score one. If this scheme is used, the asymptotic normality and optimality of the exponential scores estimate carries over to Type II censored samples in the two-sample case (Doksum (1967)).

4. Comparisons

From the considerations in Section 3, we know that for the Weibull model without any censoring, the rank exponential scores estimate $\hat{\underset{\sim}{\theta}}_R$ is asymptotically optimal. The asymptotic efficiency of the least squares type estimate $\hat{\theta}_{\sim LS}$ is $e(\hat{\theta}_{\sim LS}, \hat{\theta}_{\sim R}) = (6/\pi^2) = .61$ for all values of the Weibull parameter. To study the efficiency of the Cox estimate, we need to consider the two-sample problem. We let the parameter of interest be the ratio δ of the means of the survival distributions. When $\gamma = 1$, we find

δ	1	2	4	8	16
$e(\hat{\delta}_{COX}, \hat{\delta}_{LS})$	1.6	1.5	1.2	.83	.55
$e(\hat{\delta}_{COX}, \hat{\delta}_{R})$	1	.90	.71	.50	.33

From the results of Section 3, a qualitatively similar result should hold for $\gamma \neq 1$, Type II censoring and more general designs, but we do not have exact figures.

5. DISCUSSION

The asymptotics for the Weibull and increasing failure rate average models clearly favor the rank exponential scores estimate. However, this estimate is hard to compute and its finite sample size properties are not well known. Moreover, if we consider a different model such as the log normal model for the distribution of T_i, then the LS type estimate will be best in the case of no censoring. In this model, the optimal rank estimate would be the rank normal scores estimate.

ACKNOWLEDGEMENT

This research was partially supported by NSF grant MCS-81-92349.

REFERENCES

Aalen, O. (1978). Nonparametric inference for a family of counting processes. Annals of Statistics **6**, 701-726.

Aalen, O. (1980). A model for nonparametric regression analysis of counting processes. Proceedings, Sixth International Conference, Mathematical Statistics and Probability Theory, Wisla, Poland, Springer Lecture Notes in Statistics.

Adichie, J. (1967). Estimation of regression parameters based on rank tests. Annals of Mathematical Statistics **38**, 894-904.

Basu, A.P. (1967). On a generalized Savage statistic with applications to life testing. Annals of Mathematical Statistics **39**, 1591-1604.

Bailey, K.R. (1979). The general maximum likelihood approach to the Cox regression model. PhD dissertation, University of Chicago, Chicago, Ill.

Buckley, J. and James, I. (1979). Linear regression with censored data. Biometrika **66**, 429-436.

Cox, D.R. (1972). Regression models and life tables (with discussion). Journal of the Royal Statistical Society B **34**, 187-220.

Crowley, J. (1974). Asymptotic normality of a new nonparametric statistic for use in organ transplant studies. Journal of the American Statistical Association **69**, 1006-1011.

Doksum, K.A. (1967). Asymptotically optimal statistics in some models with increasing failure rate averages. Annals of Mathematical Statistics **38**, 1731-1739.

Doksum, K.A. (1969). Minimax results for IFRA scale alternatives. Annals of Mathematical Statistics **40**, 1778-1783.

Doksum, K.A. (1975). Measures of difference in reliability. Proceedings of the Conference on Reliability and Fault Tree Analysis, Barlow, Fussell and Singpurwalla, eds., SIAM, 427-499.

Efron, B. (1967). The two-sample problem with censored data. Proceedings of the Fifth Berkeley Symposium on Mathematical Statistics and Probability, Vol. IV. University of California Press, Berkeley, California, 831-853.

Efron, B. (1977). The efficiency of Cox's likelihood function for censored data. Journal of the American Statistical Association **72**, 557-565.

Gastwirth, J.L. (1965). Asymptotically most powerful rank tests for the two-sample problem with censored data. Annals of Mathematical Statistics **36**, 1243-1247.

Gehan, E.A. (1965). A generalized Wilcoxon test for comparing arbitrary singly-censored samples. Biometrika **52**, 203-223.

Hájek, J. and Šidák, Z. (1967). Theory of Rank Tests. Academic Press, New York.

Hodges, J. and Lehmann, E. (1963). Estimates of location based on rank tests. Annals of Mathematical Statistics **34**, 598-611.

Johnson, R.A. and Mehrotra, K.G. (1972). Locally most powerful rank tests for the two-sample problem with censored data. Annals of Mathematical Statistics **43**, 823-831.

Jureckova, J. (1971). Nonparametric estimate of regression coefficients. Annals of Mathematical Statistics **42**, 1328-1338.

Kalbfleisch, J.D. and Prentice, R.L. (1980). The Statistical Analysis of Failure Time Data. Wiley, New York.

Koul, H., Susarla, V. and Van Ryzin, J. (1981). Regression analysis with randomly right censored data. Annals of Statistics **9**, 1276-1288.

Kuczma, M. (1968). Functional Equations in a Single Variable, Polish Scientific Publishers, Warszawa.

Lehmann, E.L. (1953). The power of rank tests. Annals of Mathematical Statistics **24**, 23-43.

Louis, T.A. (1981). Nonparametric analysis of an accelerated failure time model. Biometrika **68**, 381-390.

Mantel, N. (1966). Evaluation of survival data and two new rank order statistics arising in its consideration. Cancer Chemotherapy Reports **50**, 163-170.

Miller, R.G. (1976). Least squares regression with censored data. Biometrika **63**, 449-464.

Nabeya, Seiji (1974). On the functional equation f(p+qx+rf(x)) = a+bx+rf(x). Aequationes Mathematicae **11**, 199-211.

Oakes, D. (1977). The asymptotic information in censored survival data. Biometrika **64**, 441-448.

Peto, R. and Peto, J. (1972). Asymptotically efficient rank invariant test procedures. Journal of the Royal Statistical Society A **135**, 185-206.

Prentice, R.L. (1978). Linear rank tests with right censored data. Biometrika **65**, 167-179.

Rao, V.V.R., Savage, I.R. and Sobel, M. (1960). Contributions to the theory of rank order statistics: The two-sample censored case. Annals of Mathematical Statistics **31**, 415-426.

Savage, I.R. (1956). Contributions to the theory of rank order statistics - the two-sample case. Annals of Mathematical Statistics **27**, 590-615.

Tsiatis, A. (1981). A large sample study of Cox's regression model. Annals of Statistics **9**, 93-108.

MULTI-STEP ESTIMATION OF REGRESSION COEFFICIENTS IN A LINEAR MODEL WITH CENSORED SURVIVAL DATA

Hira L. Koul
Michigan State University

V. Susarla
SUNY - Binghamton

John Van Ryzin
Columbia University and Brookhaven National Laboratory

1. Introduction

This paper introduces a multi-step procedure for estimating the regression coefficients in a linear model when the dependent variable of interest is a randomly right-censored transform of survival, i.e., log lifetime. The procedure is closely related to that introduced by Buckley and James (1979). Using large sample properties developed by the authors (1981), asymptotic large sample consistency and normality are seen to hold for each iterate of the original estimator. A limited simulation study examines the small sample behavior of the procedure.

The linear regression model considered is:

$$X_i = \alpha + \beta x_i + \varepsilon_i, \ i = 1,\dots,n \quad , \tag{1}$$

where $\{x_i\}$ are the known independent (design) variables, $\{\varepsilon_i\}$ are independent, identically distributed (i.i.d.) error variables with an unknown distribution function F with $E(\varepsilon_1) = 0$, $Var(\varepsilon_1) = \sigma^2$, and (α, β) are the parameters of interest. There exists an extensive literature on inference for α and β based on observing the X_i, but only recently has much work been done on estimating α and β when the X_i's are right-censored. For a discussion of some recent results, see Miller (1976,1981), Buckley and James (1979), and Koul, Susarla and Van Ryzin (1981, 1982). The importance of such a problem in survival data analysis where the X_i's are survival times, or transforms thereof such as log lifetimes, has been pointed out in the above references. Typically right-censored data with follow-up times Y_i can be represented as

$$Z_i = \min\{X_i, Y_i\}$$

and

$$\delta_i = \begin{cases} 1 & \text{if } X_i \leq Y_i \quad \text{(uncensored lifetime)} \\ 0 & \text{if } X_i > Y_i \quad \text{(censored lifetime)}, \end{cases}$$

where, here and throughout, $i = 1, \ldots, n$. The paper by Miller (1976) considers the situation where $Y_i = \alpha + \beta x_i + \varepsilon_i'$, where $\{\varepsilon_i'\}$ are i.i.d., independent of $\{\varepsilon_i\}$, and proposes estimators of α and β via the method of Kaplan-Meier (1958) least squares. That is, he suggests minimizing the sum of squares of residuals with weights assigned to the summands in the sum of squares according to the Kaplan-Meier estimator of F based on the residuals. Such a solution leads to an iterative procedure for the estimators of α and β which Miller shows may not converge and could lead to inconsistent estimators of α and β unless the above assumption holds, namely that the Y_i's have means following the regression line. To overcome these possible inconsistency problems, Buckley and James (1979) suggest another method, described below, which also may have the same type of convergence problem as does the Miller method, although less so. In Buckley and James, the $\{Y_i = y_i\}$ is taken as a fixed known sequence. In this paper, we present a modification of the Buckley-James procedure under the assumption that the $\{Y_i\}$ are i.i.d. random variables. The first-step estimators suggested for the α and β are those presented by Koul, Susarla and Van Ryzin (1981), which were shown to be consistent and asymptotically normal. Using these as first-step estimators, we show that the iterated subsequent-step estimators are also consistent and asymptotically normal.

The next section describes the method of Buckley and James (1979). Section 3 presents our modification of their method, while Section 4 provides the large sample consistency and asymptotic normality of the estimator. Section 5 contains some simulations of our method and that of Buckley-James and compares subsequent iterates using as first-step estimators those of Koul, Susarla and Van Ryzin (1981).

2. The Buckley-James Method

Consider the random variable

$$X_i^* = \delta_i X_i + (1 - \delta_i)\, E[X_i | X_i > y_i] \tag{2}$$

for $i = 1, \ldots, n$. Note that X_i^* is an unbiased estimator of $\alpha + \beta x_i$ for each i. Hence, with $d_i = x_i - \bar{x}$, $n\bar{x} = \sum_i x_i$, where $\sum_i$ represents sum over $i = 1, \ldots, n$, we have, $E[\sum_i d_i(X_i^* - \beta x_i)] = 0$. Therefore, a suitable analogue of the normal equation solution for estimating β is to choose that value of b such that

$$b = (\textstyle\sum_i d_i X_i^*) / \tau_x^2, \qquad \tau_x^2 = \textstyle\sum_i (x_i - \bar{x})^2 \quad .$$

However, the factor $E[X_i | X_i > y_i]$ is unknown in X_i^* and therefore b cannot be calculated directly from the data. Buckley and James (1979) suggest estimating this expectation by using the product limit estimator $\hat{F}_{0,b}(\cdot)$ based on the censored residuals $(\delta_i, Z_i - bx_i)$. Then

(3) $$E[X_i | X_i > y_i] = bx_i + E[X_i - bx_i | X_i - bx_i > y_i - bx_i]$$

can be estimated by

(4) $$\overline{X}_i(b) = bx_i + \sum_{k \varepsilon A_i} (y_k - bx_k)\ v_k(b) / \hat{F}(y_i - bx_i) \quad ,$$

where $v_k(b)$ is the jump under $\hat{F}_{0,b}$ at all uncensored $X_k - bx_k$ and A_i is the set of all k such that $X_k - bx_k > y_i - bx_i$. Note that the y_i's are fixed. Substituting (4) as an estimator of (3) into b gives the equation

(5) $$b = \sum_i [\delta_i d_i Z_i + (1 - \delta_i)\ \overline{X}_i(b)] / \tau_x^2 \quad .$$

Buckley and James suggest solving for b in (5) iteratively using $b_0 = \sum_i(\delta_i d_i Z_i)/\tau_x^2$ as the initial estimator of β. They also provide some simulation results for their estimator and a heuristic discussion of its large sample properties.

3. A Modification of the Buckley-James Method

In our modification of Buckley and James (1979), we assume $\{Y_i\}$ to be i.i.d. random variables with some continuous distribution G and that $\{Y_i\}$ is independent of $\{\varepsilon_i\}$. Consider again the relation of (1) written as

(6) $$X_i^* = \delta_i X_i + (1 - \delta_i)\ E[X_i | \delta_i = 0] \quad .$$

It is easy to see that $E[X_i^*] = \alpha + \beta x_i$ for all i, and the X_i^* follow the regression line of interest. Thus, the Buckley-James observation holds when the $\{Y_i\}$ are random provided the $\{X_i\}$ and $\{Y_i\}$ are independent. However, the $\{Y_i\}$ could be dependent within themselves. Thus, natural estimators of β and α

based on the X_i^* in (6) are

$$\tilde{\beta} = (\textstyle\sum_i d_i X_i^*) / \tau_x^2 \quad , \quad \tilde{\alpha} = n^{-1} \textstyle\sum_i X_i^* - \tilde{\beta}\,\bar{x} \quad . \tag{7}$$

The X_i^* in (6) themselves cannot be used directly since the $E[X_i | \delta_i = 0]$ are unknown. To overcome this difficulty we will estimate $E[X_i | \delta_i = 0]$ based on a preliminary estimate of β, say, β^*.

Note that the definition of conditional expectation implies

$$E[X_i | \delta_i = 0] = \frac{\int sG(s)\, dF(s - \alpha - \beta x_i)}{\int G(s)\, dF(s - \alpha - \beta x_i)} = \frac{P_{i,1}}{P_{i,0}} \quad , \tag{8}$$

where $P_{i,j}$, $j = 0,1$ are defined as indicated. Therefore, if a first-step estimator (α^*, β^*) is available, one can estimate $1 - F$ by $1 - F^*$ given by the Kaplan-Meier (1958) estimator based on the censored residuals $(\delta_i, Z_i - \alpha^* - \beta^* x_i)$ and $1 - G$ can be estimated by the Kaplan-Meier estimator $1 - G^*$ using $(1 - \delta_i, Z_i)$; that is, we treat the lifetimes as censoring the follow-up times. Thus, $P_{i,j}$ for $j = 0,1$ can be estimated by

$$P_{i,j}^* = \int s^j\, G^*(s)\, dF^*(s - \alpha^* - \beta^* x_i) \quad , \tag{9}$$

for $i = 1,\ldots,n$ and $j = 0,1$. Substituting (9) into (7) and (8) we have a second-step estimator $\hat{\beta}$ of β given by

$$\hat{\beta} = \textstyle\sum_i d_i \{\delta_i Z_i + (1 - \delta_i)\, P_{i,1}^* / P_{i,0}^*\} / \tau_x^2 \quad . \tag{10}$$

The second-step estimator of α resulting from this procedure would be

(11) $$\hat{\alpha} = n^{-1} \sum_i \{\delta_i Z_i + (1-\delta_i)\, P^*_{i,1} / P^*_{i,0}\} - \hat{\beta}\,\bar{x} \quad .$$

Given these second-step estimators, one could then repeat this process to obtain a third-step estimator. Multi-step estimators could be derived by continuing this procedure. In the next section, we show that starting with the consistent and asymptotically normal estimators of (α,β), the second-step estimator (and hence subsequent-step) estimators are also consistent and asymptotically normal. Section 5 investigates by simulation the change in the estimators over the early steps and compares them with the Buckley-James estimators.

4. Some (Intuitive) Large Sample Properties of $\hat{\beta}$

Consider the second-step estimator $\hat{\beta}$ for β given by (10). If $\{P^*_{i,j}\}$ are consistent estimators of $\{P_{i,j}\}$, then $\hat{\beta}$ can be expected to be a consistent estimator of β. Therefore, we consider the behavior of $P^*_{i,j}$ under i.i.d. censoring with distribution G.

Under i.i.d. censoring and certain other regularity conditions β^* can be chosen to be a consistent estimator of β (see Koul, Susarla and Van Ryzin (1981)). In fact, β^* can be taken such that $\tau_x(\beta^* - \beta) \overset{d}{\to} U \sim N(0,\sigma^2)$, where $\overset{d}{\to}$ stands for convergence in distribution as $n \to \infty$ and $N(0,\sigma^2)$ stands for a normal distribution with mean zero and variance σ^2.

We can express the difference $\hat{P}_{i,0} - P_{i,0}$ as

(12) $$\hat{P}_{i,0} - P_{i,0} = \int \{G^*(s+\alpha^*+\beta^* x_i) - G(s+\alpha^*+\beta^* x_i)\}\, dF^*(s)$$
$$+ \int G(s+\alpha^*+\beta^* x_i)\, d(F^* - F)\,(s) \quad .$$

Note that the first term on the right-hand side of (12) can be bounded by $\sup_s |G^*(s) - G(s)|$ which converges to zero with probability one under quite general conditions due to the recent results of Földes and Rejtö (1981). The same situation exists for the second term on the right-hand side of (12) which is bounded by $\sup_s |F^*(s) - F(s)|$. Since these convergent bounds are independent of i, equation (12) yields $\lim_n \sup_i |P^*_{i,0} - P_{i,0}| = 0$ with probability one. Similarly, it can be expected that $\lim_n \sup_i |P^*_{i,1} - P_{i,1}| = 0$ with probability one, under quite general conditions. For example, using results similar to those of Susarla and Van Ryzin (1980) for estimating the mean, it would suffice to have the $\{x_i\}$ be bounded and the G having a heavier tail than F. Hence, from these results, we expect $\hat{\beta}$ of (10) to be a strongly consistent estimator of β.

Note that there is no need to have an estimator of α to implement (10) to find $\hat{\beta}$. This is similar to the situation noted by Buckley and James (1979) in their estimate of β. Thus, in the remainder of the arguments we take $\alpha = \alpha^* = 0$ for simplicity of notation.

To study the asymptotic normality of the second-step $\hat{\beta}$, we consider the random variable $\tau_x(\hat{\beta}-\beta)$. In all statements which follow, by $o_p(1)$ we mean a term which converges to zero in probability as $n \to \infty$. Consider now

$$\tau_x(\hat{\beta}-\beta) = \tau_x^{-1} \sum_i \left\{ d_i(1-\delta_i) \left(\frac{\int (s+\beta^* x_i)\, G^*(s+\beta^* x_i)\, dF^*(s)}{\int G^*(s+\beta^* x_i)\, dF^*(s)} - E[X_i | 1-\delta_i] \right) \right\} \tag{13}$$

$$+ \tau_x^{-1} \sum_i d_i(\delta_i Z_i - E[\delta_i Z_i]) = I + II \;.$$

Since II is easily seen to converge in distribution as a sum of independent random variables provided $\tau_x^2 \to \infty$ as $n \to \infty$, we concentrate our efforts on obtaining an approximation (in probability) of I. For the following details, let

$F_i(s) = F(s - \beta x_i)$, $a_i = \int sG(s)\, dF_i(s)$, and $b_i = \int (1 - F_i(s))\, dG(s)$ and for $i = 1,\ldots,n$, $F_i^*(s) = F^*(s - \beta^* x_i)$. Note that $F_i^*(s)$ is the Kaplan-Meier estimator based only on $\{\delta_i, z_i - \beta^* x_i)$. Now write I as:

$$I = \tau_x^{-1} \sum_i \left\{ d_i(1-\delta_i) \left(\frac{\int sG^* \, dF_i^*}{\int G^* \, dF_i^*} - \frac{\int sG \, dF_i}{\int G^* \, dF_i^*} \right) \right\}$$

(14)
$$+ \tau_x^{-1} \sum_i \left\{ d_i(1-\delta_i) \left(\frac{\int sG \, dF_i}{\int G^* \, dF_i^*} - \frac{\int sG \, dF_i}{\int G \, dF_i} \right) \right\}$$

$$+ \tau_x^{-1} \sum_i \left\{ d_i(1-\delta_i) \left(\frac{\int sG \, dF_i}{\int G \, dF_i} - E[X_i | 1-\delta_i] \right) \right\}$$

$$= I_1 + I_2 + I_3 \quad .$$

Since I_3 is a sum of independent random variables, it can be shown to be asymptotically normal. Thus, consider I_1 and I_2.

We first deal with I_2 which can be rewritten as

$$I_2 = \tau_x^{-1} \sum_i d_i(1-\delta_i) \left\{ \int sG \, dF_i \right\} \frac{[\int G \, dF_i - \int G^* \, dF_i^*]}{(\int G \, dF_i)(\int G^* \, dF_i^*)}$$

(15)
$$= \tau_x^{-1} \sum_i d_i(1-\delta_i) \frac{\int sG \, dF_i}{(\int G \, dF_i)^2} \left\{ \int G \, dF_i - \int G^* \, dF_i^* \right\} + o_p(1)$$

$$= \tau_x^{-1} \sum_i \frac{a_i}{b_i^2} d_i(1-\delta_i) \left\{ \int (G - G^*)\, dF_i^* + \int G \, d(F_i - F_i^*) \right\} + o_p(1)$$

$$= \tau_x^{-1} \sum_i \frac{a_i}{b_i^2} d_i(1-\delta_i) \left\{ \int (G - G^*)\, dF_i + \int (F_i - F_i^*)\, dG \right\} + o_p(1) \quad .$$

Since the term in braces in the last line of (15) is approximately centered, replace $(1-\delta_i)$ by its expectation b_i to obtain

$$(16) \qquad I_2 = \tau_x^{-1} \sum_i d_i \frac{a_i}{b_i} \{ \int (G - G^*)\, dF_i + \int (F_i - F_i^*)\, dG\} + o_p(1) \quad .$$

The first term of (16) can be treated by an approximation as in Koul, Susarla and Van Ryzin (1981). This involves first approximating the term by a U-statistic, and then reducing it to a sum of independent centered random variables, denoted by $A_1 + \cdots + A_n$.

To approximate the term $\sum_i d_i(a_i/b_i) \int (F_i - F_i^*)dG$ in (16), we have to study the process $\{|F_i^*(s) - F_i(s)| - \infty < s < \infty\}$, for $i = 1,\ldots,n$, more carefully. Recall here that $F_i(s) = F(s-\beta x_i)$. A Taylor expansion yields

$$F_i^*(s) - F_i(s) = \exp\{\ln F_i^*(s)\} - \exp\{\ln F_i(s)\}$$

$$(17) \qquad = \sum_j \frac{\delta_j[Z_j - \beta x_j < s - \beta x_i + (\beta^* - \beta)(x_j - x_i)]}{1 + \sum_k [Z_k - \beta x_k > Z_j - \beta x_j + (\beta^* - \beta)(x_j - x_k)]} - \ln F_i(s) + o_p(1) \ ,$$

where $o_p(1)$ is independent of i and $[S]$ is the indicator function of the set S. The random variable in (17) can further be approximated by

$$\frac{1}{n} \sum_j \delta_j \frac{[Z_j - \beta x_j < s - \beta x_i + (\beta^* - \beta)(x_j - x_i)]}{n^{-1} \sum_k [Z_k - \beta x_k > Z_j - \beta x_j] + n^{-1} \sum_k (x_j - x_k)\, f_k(Z_j - \beta x_j)(\beta^* - \beta)}$$

$$- \ln F_i(s) + o_p(1) \quad ,$$

where f_k is the density of $Z_k - \beta x_k$. The factor

$|\beta^*-\beta|\,|x_j| < \tau_x|\beta^*-\beta|(\tau_x^{-1}\max_i|x_i|) \xrightarrow{p} 0$ whenever $\max_i|x_i|\,\tau_x^{-1} \to 0$ and $\beta^* \xrightarrow{p} \beta$ as $n \to \infty$. Therefore,

$$(19)\quad F_i^*(s) - F_i(s) = \frac{1}{n}\sum_j \frac{\delta_j[Z_j - \beta x_j < s - \beta x_i]}{n^{-1}\sum_k[Z_k - \beta x_k > z_j - \beta x_j]} - \ln F_i(s) + o_p(1) \quad .$$

This term is similar to the terms dealt with in Koul, Susarla and Van Ryzin (1981) which can be approximated by a sum of independent centered random variables, denoted by $B_1 + \cdots + B_n$. Combining this result with that earlier for the first-term of (16), we have

$$(20)\qquad I_2 = A_1 + \cdots + A_n + B_1 + \cdots + B_n + o_p(1) \quad .$$

To treat the term I_1 in (14), rewrite I_1 as

$$(21)\qquad I_1 = \tau_x^{-1}\sum_i d_i(1-\delta_i)(b_i^*)^{-1}\{\int s(G^*dF_i^* - GdF_i)\} \quad ,$$

and noting that $\lim_n b_i^* = \lim \int G^*dF_i^* = \int GdF_i = b_i$ and that $\lim_n \sup_i \sup_s |F_i^*(s) - F_i(s)| = 0$ with probability one, we can approximate I_1 as

$$(22)\quad I_1 = \tau_x^{-1}\sum_i d_i(1-\delta_i)\, b_i^{-1}\{\int s(G^* - G)dF_i + \int s(F_i^* - F_i)dG\} + o_p(1) \ .$$

By writing $G^*(s) - G(s) = \exp\{\ln G^*(s)\} - \exp\{\ln G(s)\}$ and similarly for $F_i^* - F_i$ and doing a Taylor expansion similar to that in (17), I_1 can be approximated as a sum of independent centered random variables, denoted by $C_1 + \cdots + C_n$. This result combined with (13), (14) and (20) yield $\tau_x(\hat{\beta} - \beta) = II + (A_1 + \cdots + A_n) + (B_1 + \cdots + B_n) + (C_1 + \cdots + C_n) + I_3$, which we see is the sum of a triangular array of independent random variables, and hence by the Lindeberg-Feller central limit theorem will be asymptotically normal. Note that the conditions that $\tau_x^2 \to \infty$ as $n \to \infty$, $\max_i |x_i| / \tau_x \to 0$ and that G have heavier tails than F are required for this to hold. Thus, asymptotic normality of $\hat{\beta}$ is expected to hold provided the first-step estimator β^* is consistent. Such a consistent first-step estimator is given by Koul, Susarla and Van Ryzin (1981). That subsequent finite-step estimators of β are consistent and asymptotically normal follows inductively.

5. Some Simulation Results

Table 1 presents the results of six simulations of the multi-step procedure of this paper and that of Buckley and James (1979). In all cases, the table entries are based on 500 simulations of the situation described. All simulations are for the two sample problem with $n = 50$ observations where $x_i = 0$ for $i = 1,\ldots,25$ and $x_i = 1$ for $i = 26,\ldots,50$. The simulation sample standard errors for the simulation averages of $\hat{\alpha}$ and $\hat{\beta}$ for the 500 repetitions were in all cases $\leq .017$ for $\hat{\alpha}$ and $\leq .031$ for $\hat{\beta}$ and thus are not individually given to save space. The first-step estimates for both methods were taken as the estimator of α and β as defined in Koul, Susarla and Van Ryzin (1981) with $M_n = n$, and are denoted in the tables as method M_1. The second, third and fourth-step estimators for the estimators introduced in this paper are denoted by M_2, M_3, and M_4, respectively, while those of Buckley and James are referred to as M_2^*, M_3^* and M_4^*. Furthermore, in each case of Table 1 the error distribution for the simulation was taken as $\varepsilon_i \sim N(0,1)$. The follow-up distribution G(y) for the first five cases are exponential with mean μ and are denoted by $E(\mu)$ in Table 1 while the six case has a right-sided logistic distribution given by $1 - G(y) =$

$2e^{-y}/(1+e^{-y})$ on $[0,\infty]$ and is denoted by LOG(½).

Upon examining Table 1, it is clear that the method of this paper performed better than that of Buckley-James over the initial three iterates after the first-step. Thus, based on these limited simulations, we feel the multi-step procedure introduced in this paper holds considerable promise. Simulations for differing sample sizes and for regression situations other than the two samples are under investigation and will be presented elsewhere.

TABLE 1. Simulation of 4-step estimators based on 500 replicates.

Simulated Case	Method of Estimation	Average of Estimates (α,β)	Average Mean Square Error of Estimates (α,β)
	M_1	(-.200,.017)	(.082,.077)
$(\alpha,\beta) = (0,0)$	M_2	(-.144,.015)	(.074,.087)
Follow-up dist. =E(1)	M_3	(-.137,.015)	(.076,.096)
Average censoring:	M_4	(-.135,.015)	(.077,.100)
Sample 1 = .44	M_2^*	(-.190,.013)	(.085,.082)
Sample 2 = .44	M_3^*	(-.186,.012)	(.087,.088)
	M_4^*	(-.185,.012)	(.088,.092)
	M_1	(-.227,-.797)	(.087,.115)
$(\alpha,\beta) = (0,-1)$	M_2	(-.130,-.906)	(.067,.096)
Follow-up dist. = E(2)	M_3	(-.105,-.939)	(.066,.098)
Average censoring:	M_4	(-.096,-.949)	(.066,.099)
Sample 1 = .44	M_2^*	(-.200,-.875)	(.084,.094)
Sample 2 = .22	M_3^*	(-.193,-.892)	(.084,.094)
	M_4^*	(-.191,-.987)	(.084,.094)

Table 1 (continued)

Simulated Case	Method of Estimation	Average of Estimates (α,β)	Average Mean Square Error of Estimates (α,β)
	M_1	(.960,.760)	(.083,1.872)
$(\alpha,\beta) = (1,2)$	M_2	(1.021,1.485)	(.053,.400)
Follow-up dist. = E(20)	M_3	(.976,1.743)	(.048,.176)
Average censoring:	M_4	(.967,1.855)	(.048,.129)
Sample 1 = .18	M_2^*	(.961,1.338)	(.049,.590)
Sample 2 = .62	M_3^*	(.927,1.560)	(.051,.306)
	M_4^*	(.908,1.655)	(.053,.214)
	M_1	(.895,.234)	(.119,3.584)
$(\alpha,\beta) = (1,2)$	M_2	(1.055,1.022)	(.076,1.217)
Follow-up dist. = E(10)	M_3	(.961,.1.387)	(.058,.576)
Average censoring:	M_4	(.942,1.597)	(.058,.326)
Sample 1 = .30	M_2^*	(.911,.815)	(.068,1.692)
Sample 2 = .79	M_3^*	(.871,1.147)	(.071,.958)
	M_4^*	(.839,1.343)	(.078,.620)
	M_1	(-.008,1.708)	(.047,1.895)
$(\alpha,\beta) = (0,3)$	M_2	(.012,2.506)	(.046,.366)
Follow-up dist. =E(20)	M_3	(-.008,2.776)	(.044,.155)
Average censoring:	M_4	(-.012,2.884)	(.043,.118)
Sample 1 = .07	M_2^*	(-.012,2.276)	(.044,.662)
Sample 2 = .62	M_3^*	(-.037,2.487)	(.044,.372)
	M_4^*	(-.052,2.574)	(.045,.277)

Table 1 (continued)

Simulated Case	Method of Estimation	Average of Estimates (α,β)	Average Mean Square Error of Estimates (α,β)
	M_1	(-.248,-.790)	(.094,.113)
$(\alpha,\beta) = (0,-1)$	M_2	(-.144,-.907)	(.066,.091)
Follow-up dist. = LOG(½)	M_3	(-.115,-.945)	(.064,.093)
Average censoring:	M_4	(-.104,-.959)	(.064,.095)
Sample 1 = .50	M_2^*	(-.234,-.884)	(.094,.087)
Sample 2 = .26	M_3^*	(-.227,-.908)	(.093,.087)
	M_4^*	(-.222,-.915)	(.093,.088)

6. Concluding Remarks

This paper presents a multi-step estimator for the α and β in linear model (1) when the independent variable X_i is randomly right-censored. These estimators are modifications of the Buckley-James estimators. The large sample properties shown to hold here might be extendable to the Buckley-James case of fixed censoring or to our modification of the Buckley-James case when the Y_i are not i.i.d. (see Susarla and Van Ryzin (1979)) if in our method we replace $(1-\hat{G})$ in our formulas by the estimator one would get using the Kaplan-Meier for the non i.i.d. case as an estimator of $\lim_n\{n^{-1}\sum_i(1-G_i(t)\}$, assuming this exists. This seems worth further investigation.

We remark that everything mentioned in this paper easily extends to the multiple regression model where

$$X_i = \underset{\sim}{C}_i \underset{\sim}{\beta} + \varepsilon_i, \ i=1,\ldots,n \quad ,$$

with $\underset{\sim}{C}_i$ being the ith row of the $n \times p$ design matrix C, $\underset{\sim}{\beta}$ is the $p \times 1$ vector of regression coefficients, and G_i, Y_i, Z_i and δ_i are as above. Then, the second-step estimator of $\underset{\sim}{\beta}$ would be given by

$$\hat{\underset{\sim}{\beta}} = (C'C)^{-1} C' \hat{\underset{\sim}{X}}^* \quad , \quad \hat{\underset{\sim}{X}}^* = (\hat{X}_1^*, \ldots, \hat{X}_n^*)' \quad ;$$

with $\hat{X}_i^* = \delta_i Z_i + (1-\delta_i) P_{i,1}^* / P_{i,0}^*$ where $P_{i,j}^*$ is given by (9) with the first-step uncensored residuals $Z_i = \underset{\sim}{C}'\underset{\sim}{\beta}^*$ with $\delta_i = 1$ used to estimate $1 - F^*$ by the Kaplan-Meier method, $1 - G^*$ is estimated as before, and $\underset{\sim}{\beta}^*$ is the first-step estimate of $\underset{\sim}{\beta}$ given by (5.2) in Koul, Susarla and Van Ryzin (1981) with $M_n = n$.

Clearly, further simulations (or theoretical) studies of the convergence properties (speed, etc.) of our multi-step procedure seem warranted and are anticipated.

ACKNOWLEDGEMENTS

The research of Professors Susarla and Van Ryzin was supported by NIH grant No. 1-R01-GM28405 at Columbia University. The authors wish to thank Ms. Sonja Johansen for the programming of the simulation results of Section 5.

REFERENCES

Buckley, J. and James, I. (1979). Linear regression with censored data. Biometrika **66**, 429-436.

Földes, A. and Rejtö, L. (1980). Strong uniform consistency for nonparametric survival curve estimators from randomly censored data. Annals of Statistics **9**, 122-129.

Kaplan, E.L. and Meier, P. (1958). Nonparametric estimation from incomplete observations. Journal of the American Statistical Association **53**, 457-481.

Koul, H., Susarla, V. and Van Ryzin, J. (1981). Regression analysis with randomly right-censored data. Annals of Statistics **9**, 1276-1288.

Koul, H., Susarla, V. and Van Ryzin, J. (1982). Least squares regression analysis with censored survival data. To appear in Topics in Applied Statistics (T.W. Dwivedi, Ed.). Marcel Dekker, New York.

Miller, R.G., Jr. (1976). Least squares regression with censored data. Biometrika **63**, 449-464.

Miller, R.G., Jr. (1981). *Survival Analysis*. Wiley, New York.

Susarla, V. and Van Ryzin, J. (1979). Large sample theory for survival curve estimators under variable censoring. In *Optimization Methods in Statistics*, 475-508. Academic Press, New York

Susarla, V. and Van Ryzin, J. (1979). Large sample theory for the mean survival time from censored data. *Annals of Statistics* **8**, 1002-1016.

INVERSE GAUSSIAN REGRESSION AND ACCELERATED LIFE TESTS

Gouri K. Bhattacharyya and Arthur Fries

Department of Statistics, University of Wisconsin, Madison

1. Introduction

A parametric analysis of accelerated life test data largely depends on the model chosen for the distribution of the life time and the relation of the parameter(s) to the stress variable. In addition to the consideration of empirical fit, a life distribution derived from reasonable postulates of the underlying failure process adds credence to its statistical use. The exponential, Weibull and log-normal families have been the popular choices in the extensive literature of engineering applications of accelerated stress testing. The first two draw from the extreme value theory and have simple forms of the failure rate function while the third is capable of using the large resources of the normal theory inference results. In regard to the parameter-stress relation, some empirical engineering models, such as the Arrhenius, Eyring and inverse power law, are ordinarily used. These are cast in a common framework that makes the logarithm of the scale parameter of the life distribution linearly related to the stress. Consequently, the distribution of the log-life is in a location-scale form with a linear regression for the location. Inference procedures under these formulations are discussed in Mann, Schafer and Singpurwalla (1974), Nelson (1971) and others.

This article focuses on a versatile but not so well known life distribution, called the inverse Gaussian distribution $IG(\theta,\lambda)$, whose probability density function (pdf) is given by

$$(1) \qquad f(y;\theta,\lambda) = (2\pi\lambda^{-1}y^{3})^{-\frac{1}{2}} \exp\left\{-\frac{1}{2}\lambda y^{-1}(y\theta^{-1}-1)^{2}\right\}, \quad y>0,\ \theta>0,\ \lambda>0 .$$

Its mean and variance are θ and θ^3/λ, respectively. Although it is not a location-scale family, it has the rare confluence of the three desirable features: a wide variety of shapes of the probability density curve, analytical tractability of many inferential results, and most important, its motivation from a plausible stochastic setting of the failure process. Tweedie (1957) studied the basic properties of this distribution, and an extensive literature has evolved in the last decade concerning mainly the one- and two-sample inference procedures and applications. A comprehensive survey is given by Folks and Chhikara (1978).

In the context of accelerated life tests, we refer to the physical basis of the inverse Gaussian distribution in Section 2 in order to formulate a plausible stochastic relation of the failure time to the intensity of stress x. The resulting regression model has the reciprocal-linear form $\theta^{-1} = \alpha + \beta x$ which is explored in the subsequent sections from the aspects of maximum likelihood and least squares estimation. The only previous work in this area is due to Davis (1977) who considers the traditional linear regression of the mean $\theta = \alpha + \beta x$ and assumes λ constant or proportional to θ^2. These formulations are somewhat artificial when viewed in the background of a Wiener process, and except for the uninteresting special case $\alpha = 0$, they lack the analytical advantage of our reciprocal-linear formulation.

2. The Reciprocal-Linear Regression Model

Along the ideas behind the development of the Birnbaum-Saunders (1969) fatigue life distribution, the genesis of the inverse Gaussian distribution can be cast in the context of fatigue growth in a material. Specifically, consider that a material fails when its accumulated fatigue, or depletion of strength exceeds a critical amount $\omega > 0$. Assume that the fatigue growth takes place over time according to a Wiener process with drift $\mu > 0$, and let δ^2 denote the diffusion constant of the process. Then the time to failure (y), alternatively called the first passage time through ω, has the inverse Gaussian distribution $IG(\theta,\lambda)$ with $\theta = \omega\mu^{-1}$ and $\lambda = \omega^2\delta^{-2}$ (cf. Cox and Miller 1965, p. 221).

For a stochastic relation of y to the intensity of stress x, we note that the parameter μ is the most obvious candidate to have a direct relation to x because it measures the mean fatigue growth per unit of time. A linear form $\mu = \alpha + \beta x$ with the natural positivity constraint $\beta \geq 0$ is a simple choice of the relation. On the other hand, the quantities δ and ω correspond, respectively, to the internal variability of the material and the critical damage that identifies a failure. It is therefore reasonable to assume that these are constants unrelated to x. Referring to the pdf (1) and absorbing ω into the parameters α and β, we then have the reciprocal-linear regression structure

$$\theta^{-1} = \alpha + \beta x \ , \ \lambda = \text{constant} > 0$$

(2)

$$\alpha \geq 0, \ \beta \geq 0, \ \alpha + \beta > 0, \ x > 0 \ .$$

The constancy of λ is analogous to the homoscedasticity assumption in the normal theory linear model. The positivity constraints in (2) are demanded by the fact that the pdf (1) is defined for $0 < \theta < \infty$. In a practical setting, we only require that $\alpha + \beta x > 0$ on a finite interval of x which corresponds to the admissible range of the stress. For a concrete discussion we assume that the origin is taken at the lower end point of this interval so $\alpha \geq 0$.

We now consider the observations (x_i, y_i), $i = 1,\ldots,n$ from n runs of an accelerated life test experiment where y_i denotes the failure time corresponding to the stress setting x_i. The random variables $y_1,\ldots,y_n$ are independent and y_i is distributed as $IG(\theta_i,\lambda)$, with $\theta_i^{-1} = \alpha + \beta x_i$.

Referring to (1), the log-likelihood function is given by

$$\ell = -\frac{n}{2}\log(2\pi\lambda^{-1}) - \frac{\lambda}{2}\sum_{i=1}^{n} y_i^{-1}\left\{y_i(\alpha+\beta x_i)-1\right\}^2 - \frac{3}{2}\sum_{i=1}^{n}\log y_i$$

(3)

$$= c(\alpha,\beta,\lambda) - \left\{\frac{\lambda\alpha^2}{2}\sum y_i + \lambda\alpha\beta\sum x_i y_i + \frac{\lambda\beta^2}{2}\sum x_i^2 y_i + \frac{\lambda}{2}\sum y_i^{-1}\right\} - \frac{3}{2}\sum\log y_i ,$$

which is defined on the restricted parameter space

$$\Omega = \{(\alpha,\beta,\lambda):\ \alpha \geq 0,\ \beta \geq 0,\ \alpha+\beta > 0,\ \lambda > 0\} .$$

The second form of (3) shows that we have an exponential family of distributions with four-dimensional sufficient statistics. However, the natural parameter space being only three-dimensional, the standard theories of inference for the exponential families do not readily apply.

3. Maximum Likelihood Estimation

For the moment we disregard the restricted form of the parameter space and consider maximization of ℓ with respect to $\underset{\sim}{\psi} = (\alpha,\beta,\lambda)'$. The first expression in (3) yields the likelihood equations

$$\partial\ell/\partial\alpha = \lambda\sum_{i=1}^{n}\{1-(\alpha+\beta x_i)y_i\} = 0$$

(4)

$$\partial\ell/\partial\beta = \lambda\sum_{i=1}^{n} x_i\{1-(\alpha+\beta x_i)y_i\} = 0$$

$$\partial\ell/\partial\lambda = \frac{1}{2}n\lambda^{-1} - \frac{1}{2}\sum_{i=1}^{n} y_i^{-1}\{1-(\alpha+\beta x_i)y_i\}^2 = 0 ,$$

We introduce convenient notation for the basic statistics:

$$(5)\qquad V_j = n^{-1}\sum_{i=1}^{n} y_i x_i^j \ , \ j=0,1,2$$

$$\bar{x} = n^{-1}\sum_{i=1}^{n} x_i \ , \quad \bar{y} = n^{-1}\sum_{i=1}^{n} y_i \ , \quad R = n^{-1}\sum_{i=1}^{n} y_i^{-1} \ .$$

The first two equations in (4) simplify to

$$\alpha V_0 + \beta V_1 = 1 \ , \quad \alpha V_1 + \beta V_2 = \bar{x} \quad ,$$

which are linear in the parameters as are the corresponding likelihood equations under the usual normal theory linear regression model. Interestingly however, the coefficients on the left side are random variables, each a linear function of $\underset{\sim}{y}$, while the terms on the right side are nonrandom. In the normal regression case, the situation is reversed. The likelihood equations yield the unique root $\hat{\underset{\sim}{\psi}}_L = (\hat{\alpha}_L, \hat{\beta}_L, \hat{\lambda}_L)'$ given by

$$\hat{\alpha}_L = (V_2 - \bar{x}V_1)\, D^{-1} = \bar{y}^{-1}(1 - V_1\hat{\beta}_L)$$

$$(6)\qquad \hat{\beta}_L = (\bar{x}V_0 - V_1)\, D^{-1} = -(nD)^{-1}\sum_{i=1}^{n} (x_i - \bar{x})(y_i - \bar{y})$$

$$\hat{\lambda}_L^{-1} = n^{-1}\sum_{i=1}^{n} (y_i^{-1} - \hat{\alpha}_L - \hat{\beta}_L x_i) = R - \hat{\alpha}_L - \hat{\beta}_L \bar{x} \quad ,$$

where $D \equiv V_0V_2 - V_1^2 > 0$ with probability 1 by the Cauchy-Schwarz inequality. Observe that $\hat{\beta}_L$ involves the usual covariance term in its numerator, but its denominator is quadratic in $\underset{\sim}{y}$. Further

$$-\left.\frac{\partial^2 \ell}{\partial \underset{\sim}{\psi} \partial \underset{\sim}{\psi}'}\right|_{\hat{\underset{\sim}{\psi}}_L} = n\hat{\lambda}_L \begin{bmatrix} V_0 & V_1 & 0 \\ V_1 & V_2 & 0 \\ 0 & 0 & \frac{1}{2}\hat{\lambda}_L^{-3} \end{bmatrix} .$$

Since this is positive definite, $\hat{\underset{\sim}{\psi}}_L$ locates the unique maximum of ℓ. We will call $\hat{\underset{\sim}{\psi}}_L$ the maximum likelihood root estimator (MLRE).

In order to obtain the maximum likelihood estimator (MLE), one needs to examine whether or not the root lies in the parameter space Ω. The last equation in (4) shows that $\hat{\lambda}_L > 0$. Also from the reduced versions of the first two, it is clear that at most one of $\hat{\alpha}_L$ and $\hat{\beta}_L$ can be negative. Thus, a violation of the constraints can occur in no more than one component of $\hat{\underset{\sim}{\psi}}_L$. It is easy to construct examples where one of $\hat{\alpha}_L$ and $\hat{\beta}_L$ can indeed be negative. In such cases, a search for the MLE requires a maximization of ℓ on the boundaries $\alpha = 0$ and $\beta = 0$. Substituting $\alpha = 0$ in (3), we find that ℓ is maximized at $\hat{\beta}_a = \bar{x}V_2^{-1}$, $\hat{\lambda}_a^{-1} = R - \bar{x}^2 V_2^{-1}$, and its maximum value is $\ell_a = c - n/2 + (n/2) \log \hat{\lambda}_a$ where c is a function of $\underset{\sim}{y}$. Similarly, with $\beta = 0$, the maximum value of ℓ is $\ell_b =$ $c - n/2 + (n/2) \log \hat{\lambda}_b$ which is attained at $\hat{\alpha}_b = V_0^{-1}$ and $\hat{\lambda}_b^{-1} = R - V_0^{-1}$. In either case, the maximizing solutions are positive and $\hat{\lambda}_{\bullet}^{-1}$ has the form $R - \hat{\alpha}_{\bullet} - \hat{\beta}_{\bullet}\bar{x}$. Finally, a comparison of these maximized likelihoods shows that $\ell_a \geq \ell_b$ if and only if $\bar{x}^2 V_0 \geq V_2$. Collecting these results together, a formal characterization of the MLE $\hat{\underset{\sim}{\psi}}$ can be stated as follows:

$$\begin{aligned} (\hat{\alpha},\hat{\beta}) &= (\hat{\alpha}_L,\hat{\beta}_L) && \text{if } V_1 < \min(\bar{x}V_0, \bar{x}^{-1}V_2) \\ &= (0, \bar{x}V_2^{-1}) && \text{if } \bar{x}^{-1}V_2 \leq V_1 < \bar{x}V_0 \\ &= (V_0^{-1}, 0) && \text{if } \bar{x}V_0 \leq V_1 < \bar{x}^{-1}V_2 \\ \hat{\lambda}^{-1} &= R - \hat{\alpha} - \hat{\beta}\bar{x} \ , \end{aligned} \tag{7}$$

where $\hat{\alpha}_L$ and $\hat{\beta}_L$ are given in (6). Note that if the root $\hat{\alpha}_L < 0$, the MLE of α gets pulled to zero, as one would anticipate. However, at the same time, $\hat{\beta}$ changes its functional form. The situation is similar when $\hat{\beta}_L < 0$.

Although the inverse Gaussian is not a location-scale parameter family, certain equivariance properties hold for the MLE's under scale changes in x or y, and location change in x. To describe these, we refer to the MLE's as functions of $\underset{\sim}{x}$ and $\underset{\sim}{y}$ by writing, for instance, $\hat{\alpha}(\underset{\sim}{x},\underset{\sim}{y})$ in place of $\hat{\alpha}$, and consider first the effects of scalar multiplications. For constants $d_1 > 0$ and $d_2 > 0$, the following relations hold

$$\hat{\alpha}(d_1\underset{\sim}{x},\ d_2\underset{\sim}{y}) = d_2^{-1}\ \hat{\alpha}(\underset{\sim}{x},\underset{\sim}{y})$$

$$\hat{\beta}(d_1\underset{\sim}{x},\ d_2\underset{\sim}{y}) = d_1^{-1}\ d_2^{-1}\ \hat{\beta}(\underset{\sim}{x},\underset{\sim}{y})$$

$$\hat{\lambda}(d_1\underset{\sim}{x},\ d_2\underset{\sim}{y}) = d_2\ \hat{\lambda}(\underset{\sim}{x},\underset{\sim}{y}) \quad .$$

Also, the effect of a translation of $\underset{\sim}{x}$ on the MLRE's is readily apparent. If $\underset{\sim}{x}$ is changed to $\underset{\sim}{x} + x_0\underset{\sim}{1}$ where $x_0 > -\min(x_1,\ldots,x_n)$, then $\hat{\alpha}_L$ changes to $\hat{\alpha}_L - x_0\hat{\beta}_L$ while $\hat{\beta}_L$ remains unchanged. However, $\hat{\alpha}$ and $\hat{\beta}$ could change their functional forms. For instance, if with the original $\underset{\sim}{x}$, $\hat{\alpha} > 0$, $\hat{\beta} > 0$ and if $x_0 > \hat{\alpha}\hat{\beta}^{-1}$, then we have $\hat{\beta} = (\bar{x}V_0 - V_1)D^{-1}$, whereas with $\underset{\sim}{x}$ translated to $\underset{\sim}{x} + x_0\underset{\sim}{1}$, the new $\hat{\alpha}$ is 0 and the new $\hat{\beta}$ is $(\bar{x} + x_0)(V_2 + 2x_0V_1 + x_0^2V_0)^{-1}$. The MLE $\hat{\lambda}$ changes accordingly. These properties help relate the MLE's under different choices of scales for the x and y variables.

4. Asymptotic Theory

Strong consistency and asymptotic normality of the maximum likelihood estimators are established in this section under some mild conditions on the limiting behavior of the design points. Letting $F_n(x) = \#\{x_i \le x,\ i=1,\ldots,n\}/n$, we henceforth assume that, as $n \to \infty$, F_n converges weakly to a proper distribution function F on $(0,\infty)$. For brevity, we will only treat the case when the true parameter point $\underset{\sim}{\psi}$ is in the interior of Ω; that is, neither α nor β is 0. Unless specified otherwise, all limits are taken as $n \to \infty$. We introduce the notation

$$\tau_j(n) = \int_0^\infty x^j(\alpha+\beta x)^{-1}\, dF_n \;, \quad \tau_j = \int_0^\infty x^j\,(\alpha+\beta x)^{-1}\, dF$$

(8)

$$c_j(n) = \int_0^\infty x^j\, dF_n \;, \quad c_j = \int_0^\infty x^j\, dF \;, \quad j = 0,1,2 \quad .$$

LEMMA. Assume that x^2 is uniformly integrable in F_n. Then R converges almost surely (a.s.) to $\alpha+\beta c_1+\lambda^{-1}$, and V_j to τ_j, $j=0,1,2$.

PROOF. Since $E(y_i) = (\alpha+\beta x_i)^{-1}$ and $E(y_i^{-1}) = \alpha+\beta x_i+\lambda^{-1}$ (cf. Tweedie, 1957), we have $E(V_j) = \tau_j(n)$ and $E(R) = \alpha+\beta c_1(n)+\lambda^{-1}$. The assumptions $\alpha > 0$ and $\beta > 0$ imply that $x^j(\alpha+\beta x)^{-1} \le \min\{\alpha^{-1}x^j, \beta^{-1}x^{j-1}\}$, so $\lim \tau_j(n) = \tau_j$ and $\lim E(R) = \alpha+\beta c_1+\lambda^{-1}$ by the uniform integrability of x. Noting that $\operatorname{Var}(y_i^{-1}) = \lambda^{-1}(\alpha+\beta x_i+2\lambda^{-1})$, the stated a.s. convergence of R would follow by an application of the Kolmogorov strong law once we verify that $\sum_{i=1}^\infty x_i/i^2 < \infty$. Because $c_2(n) \to c_2 < \infty$, for a given $\varepsilon > 0$ there exists an n_0 such that $n^{-1}x_n^2 \le \varepsilon$ for $n \ge n_0$. Consequently,

$$\sum_{i=1}^\infty x_i/i^2 \le \sum_{i=1}^{n_0} x_i/i^2 + \varepsilon^{1/2} \sum_{i=n_0+1}^\infty i^{-3/2} < \infty \quad ,$$

which establishes the desired result. For an application of the strong law to V_2 we require that $\sum_{i=1}^\infty x_i^4 \operatorname{Var}(y_i) i^{-2} < \infty$. This follows from the facts that $\operatorname{Var}(y_i) = \lambda^{-1}(\alpha+\beta x_i)^{-3}$ and $x^4(\alpha+\beta x)^{-3} \le \beta^{-3}x$. The treatment of V_0 and V_1 are similar, and the proof is concluded.

To show that both $\hat{\underset{\sim}{\psi}}_L$ and $\hat{\underset{\sim}{\psi}}$ are strongly consistent for $\underset{\sim}{\psi}$, we first refer to (6) and use the lemma to each component of $\hat{\underset{\sim}{\psi}}_L$. In particular, $\hat{\beta}_L \to g(\underset{\sim}{\tau})$, a.s. where

$$g(\underset{\sim}{\tau}) = (c_1\tau_0 - \tau_1)(\tau_0\tau_2 - \tau_1^2)^{-1} \quad .$$

The relations $\beta\tau_1 = 1 - \alpha\tau_0$ and $\beta\tau_2 = c_1 - \alpha\tau_1$ can be verified from the definition of τ_j, and a substitution of these yields $g(\tau) = \beta$, so $\hat{\beta}_L \to \beta$, a.s. Referring to (7) we observe that $\hat{\beta}_L \neq \hat{\beta}$ if and only if $\bar{x}V_0 - V_1 < 0$. However, using the lemma we have with probability one, $\bar{x}V_0 - V_1 \to c_1\tau_0 - \tau_1 > 0$. Thus, $\lim\sup|\hat{\beta}_L - \hat{\beta}| = 0$, with probability one, hence $\hat{\beta} \to \beta$, a.s. The proofs for $\hat{\alpha}$ and $\hat{\lambda}$ are analogous.

We now turn to the limiting distribution of the maximum likelihood estimators. In view of the asymptotic a.s. equivalence of $\hat{\underset{\sim}{\psi}}$ and $\hat{\underset{\sim}{\psi}}_L$, it suffices to consider the limiting distribution of $\hat{\underset{\sim}{\psi}}_L$. Referring to (6), the most direct approach would be to first establish the joint asymptotic normality of the V_j's and R and then use the δ-method. However, this process incurs some formidable expressions whose simplifications are quite tedious. Instead, we examine the first and second partial derivatives of the log-likelihood, and observe an interesting relation

$$n^{-1/2}\partial\ell/\partial\underset{\sim}{\psi} = \underset{\sim}{M}[n^{1/2}(\hat{\underset{\sim}{\psi}}_L - \underset{\sim}{\psi})] \tag{9}$$

where

$$\underset{\sim}{M} = \begin{bmatrix} \lambda\underset{\sim}{S} & \underset{\sim}{0} \\ \underset{\sim}{e}' & (\lambda\hat{\lambda}_L)^{-1}/2 \end{bmatrix} , \quad \underset{\sim}{S} = \begin{bmatrix} V_0 & V_1 \\ V_1 & V_2 \end{bmatrix}$$

$$\underset{\sim}{e}' = (e_1, e_2)$$

$$e_1 = -1 + \tfrac{1}{2}\{V_0(\hat{\alpha}_L + \alpha) + V_1(\hat{\beta}_L + \beta)\}$$

$$e_2 = -\bar{x} + \tfrac{1}{2}\{V_1(\hat{\alpha}_L + \alpha) + V_2(\hat{\beta}_L + \beta)\} \quad .$$

The exact relation (9) is more convenient to work with than the usual Taylor expansion of $\partial \ell / \partial \underset{\sim}{\psi}$ which involves the second derivatives evaluated at some undetermined intermediate point between $\hat{\underset{\sim}{\psi}}_L$ and $\underset{\sim}{\psi}$.

THEOREM. Assume that x^3 is uniformly integrable in F_n. Then $n^{1/2}(\hat{\underset{\sim}{\psi}}_L - \underset{\sim}{\psi})$ is asymptotically trivariate normal $N_3(\underset{\sim}{0}, \underset{\sim}{\Sigma})$ where

$$(10) \qquad \underset{\sim}{\Sigma}^{-1} = \begin{bmatrix} \lambda \underset{\sim}{\Delta} & \underset{\sim}{0} \\ \underset{\sim}{0}' & \lambda^{-2}/2 \end{bmatrix}, \quad \underset{\sim}{\Delta} = \begin{bmatrix} \tau_0 & \tau_1 \\ \tau_1 & \tau_2 \end{bmatrix} .$$

PROOF. As a consequence of the lemma, $e_i \to 0$, a.s. for $i = 1,2$, so $\underset{\sim}{M} \to \underset{\sim}{\Sigma}^{-1}$, a.s. Therefore, it suffices to show that $n^{-1/2} \partial \ell / \partial \underset{\sim}{\psi}$ has the limiting distribution $N_3(\underset{\sim}{0}, \underset{\sim}{\Sigma}^{-1})$. Referring to (4) we observe that $\partial \ell / \partial \underset{\sim}{\psi}$ is of the form $\sum_{i=1}^{n} \underset{\sim}{q}_i$ where $\underset{\sim}{q}_i$'s are independent but non-identically distributed random vectors with means $\underset{\sim}{0}$. Focusing on an arbitrary linear function $Z_n = n^{-1/2} \underset{\sim}{h}' \partial \ell / \partial \underset{\sim}{\psi}$ with $\underset{\sim}{h} \neq \underset{\sim}{0}$ we readily observe that $\lim \operatorname{Var}(Z_n) = \underset{\sim}{h}' \underset{\sim}{\Sigma}^{-1} \underset{\sim}{h} > 0$. Therefore, the Liapounov central limit theorem would apply once we establish that $\lim \zeta_n = 0$ where

$$\zeta_n = n^{-3/2} \sum_{i=1}^{n} E|\underset{\sim}{h}' \underset{\sim}{q}_i|^3 .$$

Denoting $\omega_i = y_i(\alpha + \beta x_i)$ and referring to (4), one can find constants $a_j \geq 0$ depending only on $\underset{\sim}{h}$, α, β and λ such that

$$|\underset{\sim}{h}' \underset{\sim}{q}_i|^3 \leq (1 + x_i)^3 \sum_{j=0}^{6} a_j \omega_i^{j-3} .$$

Consequently, $\zeta_n \leq n^{-1/2} \sum_{j=0}^{6} a_j \varepsilon_n$ where $\varepsilon_n = n^{-1} \sum_{i=1}^{n} (1 + x_i)^3 E(\omega_i^{j-3})$.

Using the positive and negative moments of the inverse Gaussian distribution (cf. Tweedie, 1957), it can be seen that ε_n is a fixed (not depending on n) linear combination of the terms

$$\nu_{jk}(n) = \int_0^\infty x^k(\alpha+\beta x)^{-j}\, dF_n(x)$$

with $0 \leq j,\ k \leq 3$. From the uniform integrability assumption, each of these $\nu_{jk}(n)$ has a finite limit. Hence $\lim \zeta_n = 0$ and the proof is concluded.

The limiting normal distribution of the MLE's, along with the a.s. convergence of the sample information matrix, can be used to construct large sample confidence intervals for the parameters. Specifically, the approximate variances of $\hat{\alpha}, \hat{\beta}$ and $\hat{\lambda}$ are $(n\hat{\lambda}D)^{-1}V_2$, $(n\hat{\lambda}D)^{-1}V_0$ and $2n^{-1}\hat{\lambda}^2$, respectively. Also, the reciprocal mean failure time $[\theta(x^*)]^{-1} = \alpha + \beta x^*$, at a specified stress level x^*, is estimated by $\hat{\alpha} + \hat{\beta}x^*$, whose approximate variance is given by $(nD\hat{\lambda})^{-1}$ $(V_2 - 2x^*\, V_1 + x^{*2}V_0)$.

5. A Least Squares Approach for Replicated Designs

Although closed form expressions were obtained in Section 3, an analytical treatment of the exact mean and variance of the MLE's does not appear to be feasible. In this section, we consider experiments with replicated observations and construct some unbiased estimators by employing a combination of the maximum likelihood and least squares principles. A similar procedure has been used by Singpurwalla (1973) when the underlying life distribution is exponential.

Consider k stress settings $x_1,\ldots,x_k$ and n_i independent failure times $(y_{i1},\ldots,y_{in_i})$ observed at x_i, $i = 1,\ldots,k$. The random variables y_{ij}, $j = 1,\ldots,$ n_i; $i = 1,\ldots,k$ are all independent with y_{ij} distributed as $IG(\theta_i,\lambda)$, and $\theta_i^{-1} = \alpha + \beta x_i$. Let

$$\bar{y}_i = n_i^{-1} \sum_{j=1}^{n_i} y_{ij} , \quad N = \sum_{i=1}^{k} n_i ,$$

(11)

$$Q = \sum_{i=1}^{k} \sum_{j=1}^{n_i} (y_{ij}^{-1} - \bar{y}_i^{-1}) .$$

Our method consists of two steps. Disregarding the regression structure, we first consider the θ_i's as free parameters in which case $\bar{y}_i$, $i=1,\ldots,k$ and Q constitute a set of complete sufficient statistics. As shown by Tweedie (1957), these statistics are all independent, $\bar{y}_i$ is distributed as $IG(\theta_i, n_i\lambda)$ and λQ is distributed as χ^2 with (N-k) degrees of freedom. Moreover $E(\bar{y}_i^{-1}) = \theta_i^{-1} + (n_i\lambda)^{-1}$ and $Var(\bar{y}_i^{-1}) = (\theta_i n_i\lambda)^{-1} + 2(n_i\lambda)^{-2}$. Defining

$$\tilde{\lambda} = (N-k)\, Q^{-1} ,$$

$$t_i = \bar{y}_i^{-1} - (n_i\tilde{\lambda})^{-1} ,$$

it then follows that t_i and $\tilde{\lambda}^{-1}$ are the uniformly minimum variance unbiased estimators of θ_i^{-1} and λ^{-1}, respectively. Focusing now on the t_i's which are, in essence, the bias-corrected reciprocal means, we have the linear model $E(t_i) = \alpha + \beta x_i$ with the covariance structure

$$Var(t_i) = (\alpha + \beta x_i)(n_i\lambda)^{-1} + 2(n_i\lambda)^{-2} [1 + (N-k)^{-1}] ,$$

$$Cov(t_i, t_{i'}) = 2[n_i n_{i'} (N-k)]^{-1} \lambda^{-2} , \quad i \neq i' .$$

With large n_i's, the covariances are negligible compared to the variances. The leading term in $Var(t_i)$ is proportional to n_i^{-1} but its dependence on $(\alpha + \beta x_i)$ is the major deterrant to an application of the usual weighted least squares. In order to get unbiased estimators we consider the particular weighted least squares which minimizes $\sum_{i=1}^{k} n_i(t_i - \alpha - \beta x_i)^2$. Letting $m_j = N^{-1} \sum_{i=1}^{k} n_i x_i^j$, $j \geq 1$, the resulting unbiased estimators are

$$\tilde{\alpha} = N^{-1} \sum_{i=1}^{k} n_i \bar{y}_i^{-1} - \tilde{\beta} m_1 + \tilde{\lambda}^{-1} k N^{-1} ,$$

$$\tilde{\beta} = (m_2 - m_1^2)^{-1} N^{-1} \sum_{i=1}^{k} (x_i - m_1)(n_i \bar{y}_i^{-1} - \tilde{\lambda}^{-1}) .$$

It can be seen that the behaviors of $\tilde{\psi} = (\tilde{\alpha}, \tilde{\beta}, \tilde{\lambda})'$ under different scalings of x and/or y are identical to those of $\hat{\psi}_L$. For the balanced design $n_1 = \cdots = n_k$, $\tilde{\beta}$ is independent of $\tilde{\lambda}$. The exact χ^2 distribution of $(N-k)\lambda\tilde{\lambda}^{-1}$ and the simple linear forms of $\tilde{\alpha}$ and $\tilde{\beta}$ also enable us to derive the exact variance and covariance expressions. Dropping terms of order N^{-2}, we obtain

$$\mathrm{Var}(\tilde{\lambda}^{-1}) = 2(N-k)^{-1} \lambda^{-2}, \quad \mathrm{Cov}(\tilde{\alpha}, \tilde{\lambda}^{-1}) = \mathrm{Cov}(\tilde{\beta}, \tilde{\lambda}^{-1}) = 0 ,$$

$$\mathrm{Var}(\tilde{\alpha}) = (N\lambda)^{-1} s_2^{-2} [\alpha m_2 s_2 + \beta m_1 s_3] ,$$

(12)

$$\mathrm{Var}(\tilde{\beta}) = (N\lambda)^{-1} s_2^{-2} [\alpha s_2 + \beta m_1^{-1} (s_2^2 + s_3)] ,$$

$$\mathrm{Cov}(\tilde{\alpha}, \tilde{\beta}) = -(N\lambda)^{-1} s_2^{-2} [\alpha m_1 s_2 + \beta s_3] ,$$

where $s_2 = m_2 - m_1^2$ and $s_3 = m_1 m_3 - m_2^2$.

The asymptotic (as $N \to \infty$, $n_i N^{-1} \to r_i$, and k fixed) properties of $\tilde{\psi}$ follow directly from the asymptotics for Q and $\bar{y}_i$, $i = 1, \ldots, k$. In particular, $\tilde{\psi}$ is strongly consistent, and $N^{1/2}(\tilde{\psi} - \psi)$ is asymptotically distributed as $N_3(0, \Gamma)$ where the entries of Γ are equal to the limits of N times the corresponding expressions given in (12).

Referring to (12) and the covariance matrix Σ given in (10), we obtain the asymptotic efficiencies (AE's) of the least squares estimators. Specifically,

$$\mathrm{AE}(\tilde{\lambda}) = 1,$$

$$\mathrm{AE}(\tilde{\alpha}) = \tau_2(\tau_2 - c_1\tau_1)^{-1}(c_2 - c_1^2)^2\{c_2(c_2 - c_1^2) + \nu(c_1c_3 - c_2^2)\}^{-1} ,$$

$$\mathrm{AE}(\tilde{\beta}) = \tau_0(\tau_2 - c_1\tau_1)^{-1}(c_2 - c_1^2)^2\{(c_2 - c_1^2) + \nu(c_1^2 - 2c_2 + c_1^{-1}c_3)\}^{-1} ,$$

where $\nu = \beta\alpha^{-1}c_1$ and $\tau_j = \alpha^{-1}\sum_{i=1}^{k} r_i x_i^{\,j}(1 + \nu c_1^{-1}x_i)^{-1}$, $c_j = \sum_{i=1}^{k} r_i x_i^{\,j}$, $j = 0,1,2,3$.

For fixed c_j's, these AE's are monotonically decreasing functions of ν. Their bounds can therefore be established by considering the limits $\beta \to 0$ and $\alpha \to 0$. As $\beta \to 0$, $\tau_j \to \alpha^{-1}c_j$ and the limiting AE's are each equal to 1. By continuity, the AE's are high when α is much larger than βc_1, i.e., when the major contribution to the reciprocal mean lifetime, at the center of the design, is due to α. As $\alpha \to 0$, $\tau_j \to \beta^{-1}c_{j-1}$ and the lower bounds are given by

$$\mathrm{AE}(\tilde{\alpha}) \geq \frac{(c_2 - c_1^2)^2}{(c_1c_3 - c_2^2)(c_1c_{-1} - 1)} , \quad \mathrm{AE}(\tilde{\beta}) \geq \frac{c_{-1}(c_2 - c_1^2)^2}{(c_1^3 - 2c_1c_2 + c_3)(c_1c_{-1} - 1)} .$$

Here, $c_{-1} = \sum_{i=1}^{k} r_i x_i^{-1}$ is well-defined since we have assumed all $x_i > 0$. When some x_i values are close to 0, the quantity c_{-1} gets large and it forces the lower bounds to become small. For example, the three point design with $(x_1, x_2, x_3) = (0.1,\ 1.0,\ 2.8)$ and $(r_1, r_2, r_3) = (.4,\ .4,\ .2)$ gives $\mathrm{AE}(\tilde{\alpha}) \geq .311$ and $\mathrm{AE}(\tilde{\beta}) \geq .679$. This drawback of the least squares estimators can be overcome by suitably translating the x-scale. For instance, in the previous example, the translation to $\underset{\sim}{x}_* = \underset{\sim}{x} + \underset{\sim}{1}$ gives the lower bounds .866 and .888 for $\mathrm{AE}(\tilde{\alpha})$ and $\mathrm{AE}(\tilde{\beta})$, respectively.

Straightforward computations show that, independent of the design, the AE of the estimated reciprocal mean lifetime at the center of the design is equal to 1, i.e., $\mathrm{AE}(\tilde{\alpha} + \tilde{\beta}c_1) = 1$. Also, for any two point design we have $\mathrm{AE}(\tilde{\alpha}) = \mathrm{AE}(\tilde{\beta})$

= 1. This last result is explained by noting that when $k = 2$, $\tilde{\alpha} = \hat{\alpha}_L + N^{-1}\tilde{\lambda}^{-1}b_1$ and $\tilde{\beta} = \hat{\beta}_L + N^{-1}\tilde{\lambda}^{-1}b_2$ where b_1 and b_2 are constants. Evidently, the least squares estimators are then asymptotically equivalent to the MLE's.

6. Example

Nelson (1971) reports data on the failure times of an insulation material in a motorette test performed at four elevated temperature settings ranging from 190^0C to 260^0C. The original goal of the experiment was to determine if the mean time to failure at the design temperature of 180^0C exceeded a specified minimum requirement. However, the 260^0C data were taken on a batch of insulation different from the batch used at the other temperature settings, and it became important to investigate whether or not the data from the two batches were consistent. Nelson establishes the inconsistency of the 260^0C data by employing a combination of graphical and analytic techniques based upon the assumptions that the log-failure times are normally distributed with a constant variance and the mean depending on temperature through the Arrhenius relationship.

For illustrative purposes, we fit an IG regression model to these data excluding the 260^0C setting. Although the physical properties of the insulation material are unknown to us, it seems reasonable to hypothesize that the wear of the insulation increases until a critical amount has disintegrated or ceased to perform adequately. Thus the assumption of an underlying IG distribution is plausible. The assumption of a constant λ is tenable since the materials used in the first three levels are known to have a common source. In the absence of a mechanistic model relating the mean life (θ) and the temperature (T), we base our choice on the following observations: (i) regressing $\bar{y}_i^{-1}$ on T_i^3 gives a value of $R^2 = 99.9\%$, (ii) consistent with the IG assumption, the sample variances raised to the power $-\frac{1}{3}$ are approximately linear in T_i^3, and (iii) the transformation to $T_i^3 - (180)^3$ is convenient in that it produces positive MLE's. Without a translation of T_i^3, the restriction $\alpha \geq 0$ imposed in the definition of Ω would not be meaningful.

Guided by these considerations, we take the distribution of failure times as $IG(\theta,\lambda)$ with $\theta^{-1} = \alpha + \beta x$ where $x = 10^{-8}[T^3 - 180^3]$. By using the results of Sections 3, 4, and 5, we calculate the maximum likelihood and least squares estimates of α, β and λ^{-1} and also their approximate standard errors. The results are given in Table 1 where the standard errors are shown in parentheses.

TABLE 1. The estimates and approximate standard errors

	Parameters		
	α	β	λ^{-1}
Maximum likelihood	.0371 (.0129)	7.3260 (.3557)	.0102 (.0026)
Least squares	.0320 (.0141)	7.4316 (.3747)	.0097 (.0026)

The MLE's of the mean lives (thousands of hours) at 190^0C, 220^0C, and 240^0C are, respectively, 8.902, 2.565 and 1.606, with the associated standard errors .094, .029 and .033. The corresponding least squares estimates are 8.863, 2.565 and 1.598, with the respective standard errors .101, .030 and .034. Both sets of estimates are comparable to the observed sample means 8.782, 2.638 and 1.581.

The MLE's obtained exclusively from the 260^0C data, specifically $\bar{y}_4$ and $\hat{\lambda}_4^{-1}$, are now compared with the estimates computed using only the first three levels. From Table 1, the estimates of the mean time to failure at the fourth level are $\hat{\theta}_4 = 1.112$ and $\tilde{\theta}_4 = 1.105$ with the respective standard errors .037 and .038. These estimates agree closely with the sample mean $\bar{y}_4 = 1.116$. However, $\hat{\lambda}_4^{-1} = .1310$ differs substantially from both $\hat{\lambda}^{-1}$ and $\tilde{\lambda}^{-1}$. We perform an exact test of the hypothesis of a common λ by using $\hat{\lambda}_4$ and $\tilde{\lambda}$. Under the null hypothesis of a common λ, $10\lambda\hat{\lambda}_4^{-1}$ has an exact $\chi^2(9)$ distribution and it is statistically independent of $27\lambda\tilde{\lambda}^{-1}$ which is distributed as $\chi^2(27)$. Thus, $10\hat{\lambda}_4^{-1}(9\tilde{\lambda}^{-1})^{-1}$ is distributed as $F(9,27)$. The observed value of this statistic is 15.01 which corresponds to a p-value <.001. In agreement with Nelson, we conclude that the two batches of insulating materials are significantly different.

ACKNOWLEDGEMENT

This work was supported by the Office of Naval Research Grant N00014-78-C-0722.

REFERENCES

Birnbaum, Z.W. and Saunders, S.C. (1969). A new family of life distributions. Journal of Applied Probability **6**, 319-327.

Cox, D.R. and Miller, H.E. (1965). The Theory of Stochastic Processes. Methuen, London.

Davis, A.S. (1977). Linear statistical inference as related to the inverse Gaussian distribution. Unpublished PhD thesis, Oklahoma State University.

Folks, J.L. and Chhikara, R.S. (1978). The inverse Gaussian distribution and its statistical application - a review. Journal of the Royal Statistical Society B **40**, 263-275.

Mann, N.R., Schafer, R.E. and Singpurwalla, N.D. (1974). Methods for Statistical Analysis of Reliability and Life Data. Wiley, New York.

Nelson, W.B. (1971). Analysis of accelerated life test data. I.E.E.E. Transactions on Electrical Insulation **EI-6**, 165-181.

Singpurwalla, N.D. (1973). Inference from accelerated life tests using Arrhenius type re-parametrizations. Technometrics **15**, 289-299.

Tweedie, M.C.K. (1957). Statistical properties of inverse Gaussian distributions. Annals of Mathematical Statistics **28**, 362-377.

TRANSFORMATION OF SURVIVAL DATA

Richard A. Johnson

University of Wisconsin

1. Introduction

It is good statistical practice to perform more than one analysis on a given data set. Normal theory methods usually provide one alternative. By employing transformations their domain of application can be greatly extended. Under normal theory, we have the advantage of simple interpretations of linear regression and interactions. Coupled with the relative ease of computation and availability of diagnostic techniques, for complete samples, normal theory has much to recommend it. Except for computational ease, the other advantages are retained in the survival analysis setting.

By transforming survival data, and then applying parametric estimation methods, we obtain an estimated survival curve which may be compared to the Kaplan-Meier (1958) estimate. In a regression setting comparisons could be made with the analysis based on the Cox (1972) proportional hazards model, or the methods of Miller (1976), Buckley and James (1979) and Koul, Susarla and Van Ryzin (1981).

2. Background and Notation

We briefly review the literature that pertains to our extensions. Box and Cox (1964) suggest the family of transformations

(1)
$$x^{(\lambda)} = \begin{cases} \dfrac{x^{\lambda}-1}{\lambda} , & \lambda \neq 0 \\ \ell n(x) , & \lambda = 0 \end{cases}$$

for improving the approximation of positive random variables to normality. They tentatively assume that $X^{(\lambda)}$ has a normal distribution for some choice of λ. Under this assumption, $X^{(\lambda)}$ has density

$$x^{\lambda-1} \frac{1}{\sqrt{2\pi}} \exp\left\{-\frac{1}{2}\left(\frac{x^{(\lambda)}-\mu}{\sigma}\right)^2\right\}$$

and a random sample leads to the log-likelihood

(2)
$$\frac{-n}{2} \ell n(2\pi) - \frac{1}{2} \sum_{j=1}^{n} \left(\frac{x_j^{(\lambda)} - \mu}{\sigma}\right)^2 + (\lambda-1) \sum_{j=1}^{n} \ell n(x_j) \quad .$$

We remark that $X^{(\lambda)}$ cannot be exactly normal, except possibly for $\lambda = 0$, since its support has a finite lower bound.

Hernandez and Johnson (1980) show that, asymptotically, selecting λ to maximize the expression (2) is equivalent to selecting λ to minimize the Kullback-Liebler information number between $g_\lambda(z)$, the true pdf of $X^{(\lambda)}$, and a normal distribution $\phi_{\mu,\sigma}(z)$. That is, the information number

(3)
$$I[g_\lambda;\phi_{\mu,\sigma}] = \int g_\lambda(z) \, \ell n \left[\frac{g_\lambda(z)}{\phi_{\mu,\sigma}(z)}\right] dz$$

is minimized by this choice of λ,μ,σ. Since $g_\lambda(z) = g((\lambda z+1)^{\frac{1}{\lambda}})(\lambda z+1)^{\frac{1}{\lambda}-1}$ where $g(\cdot)$ is the pdf of X, the information number can also be expressed as

(4)
$$I[g_\lambda;\phi_{\mu\sigma}] = \int g(x) \, \ell n \left[\frac{g(x)}{\phi_{\mu,\sigma}\left(\frac{x^{\lambda}-1}{\lambda}\right) x^{\lambda-1}}\right] dx \quad .$$

Several examples appear in Hernandez and Johnson (1980).

In Section 3, we treat transformation of survival data. Because the transformation technique has proven especially effective in a regression setting, in Section 6 we extend its domain of application to survival analysis with covariates.

Carroll (1980) and Bickel and Doksum (1981) raise some question about the sampling properties of estimators determined by the Box-Cox procedure. Recent evidence, however, indicates that predictions and tests for significance of regression parameters remain valid (see Carroll and Ruppert (1981)).

3. Survival Analysis Setting

In the survival analysis setting, the times of entry into the study are haphazard or random. We assume the arrival process, for items or persons, is independent of life length. Consequently, we model the life lengths as independent identically distributed random variables with c.d.f. G(•). The time on test for the ith person will be denoted by L_i = (current time) − (entry time). We either observe x_i = life length, or censor the test at L_i. We tentatively assume that some power transformation is normal. The likelihood is then

$$(5) \quad L(\lambda,\mu,\sigma) = \prod_{i\varepsilon F} \frac{1}{(2\pi)^{\frac{1}{2}}\sigma} e^{-\frac{1}{2\sigma^2}\left(\frac{x_i^\lambda - 1}{\lambda} - \mu\right)^2} \cdot x_i^{\lambda-1} \prod_{i\notin F} \left[1 - \Phi\left(\frac{\frac{L_i^\lambda - 1}{\lambda} - \mu}{\sigma}\right)\right]$$

where $F = \{i : x_i < L_i\}$ is the set of items that fail during the trial.

The log-likelihood can be maximized numerically over λ, μ and σ. Because of the censoring, there is not even a partial analytic solution as in the complete sample case.

EXAMPLE: [Stanford Heart Transplant Data]

We consider the first $n = 184$ patients reported in Miller and Halpern (1981). Numerical minimization of $-\ell n\, L(\lambda,\mu,\sigma)$ provides the estimates (we replaced the 0 lifelength by .5)

$$\hat{\lambda} = .0042\ , \quad \hat{\mu} = 6.3706\ , \quad \hat{\sigma} = 2.4956$$

and

$$-\ell n\, L(\hat{\lambda},\hat{\mu},\hat{\sigma}) = 859.0402\ .$$

The log-normal has $\lambda = 0$, $\hat{\mu}(0) = 6.2833$, $\hat{\sigma}(0) = 2.4452$ and $-\ell n\, L(0,\hat{\mu}(0),\hat{\sigma}(0)) =$ 859.044. Because of the near equivalence of the maximized likelihoods, it is just as reasonable to take $\ell n\, X_i$ as approximately normal. In fact, Miller and Halpern (1981) use $\ell n\, X_i$ without explanation. Figure 1 displays the graph of $-\ell n\, L(\lambda,\hat{\mu}(\lambda), \hat{\sigma}(\lambda))$ versus λ. If the usual asymptotic theory

$$-2\ \ell n\ [L(\lambda,\hat{\mu}(\lambda),\hat{\sigma}(\lambda))\ /\ L(\hat{\lambda},\hat{\mu},\hat{\sigma})] \text{ approximately } \chi_1^2$$

applies, values of λ in the interval -.09 to .10 should be considered reasonable choices.

It is also possible to estimate the survival function. Proceeding as if $X^{(\hat{\lambda})}$ has a normal distribution with mean $\hat{\mu}$ and variance $\hat{\sigma}^2$, we consider the survival estimate

$$\text{(6)} \qquad \hat{S}(x) = \widehat{P[X > x]} = 1 - \Phi\left(\frac{\frac{x^{\hat{\lambda}}-1}{\hat{\lambda}} - \hat{\mu}}{\hat{\sigma}}\right)\ .$$

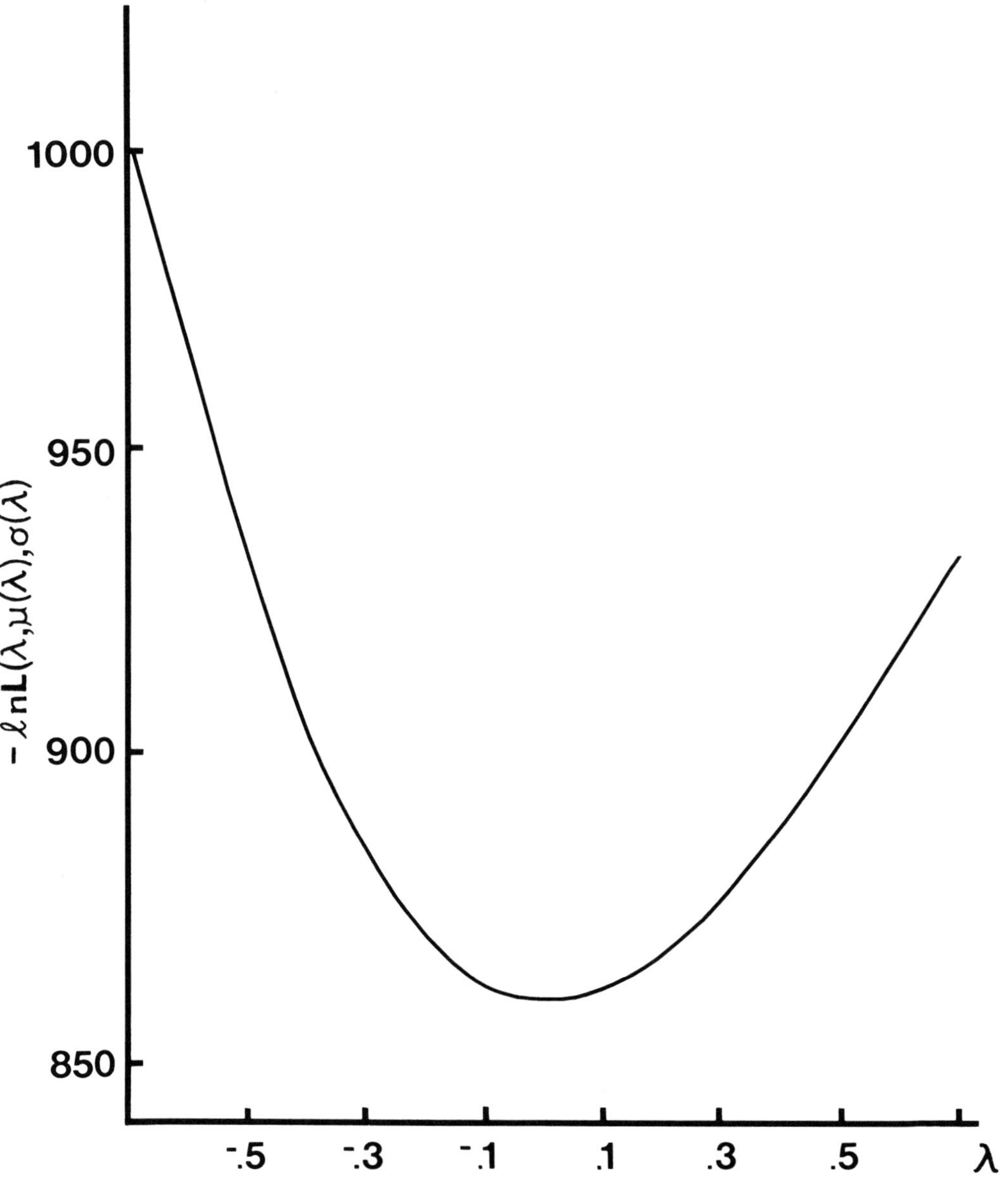

FIGURE 1: The negative of the partially maximized log-likelihood.

Figure 2 displays the estimated survival function (6), along with the Kaplan-Meier estimate, for the heart transplant data.

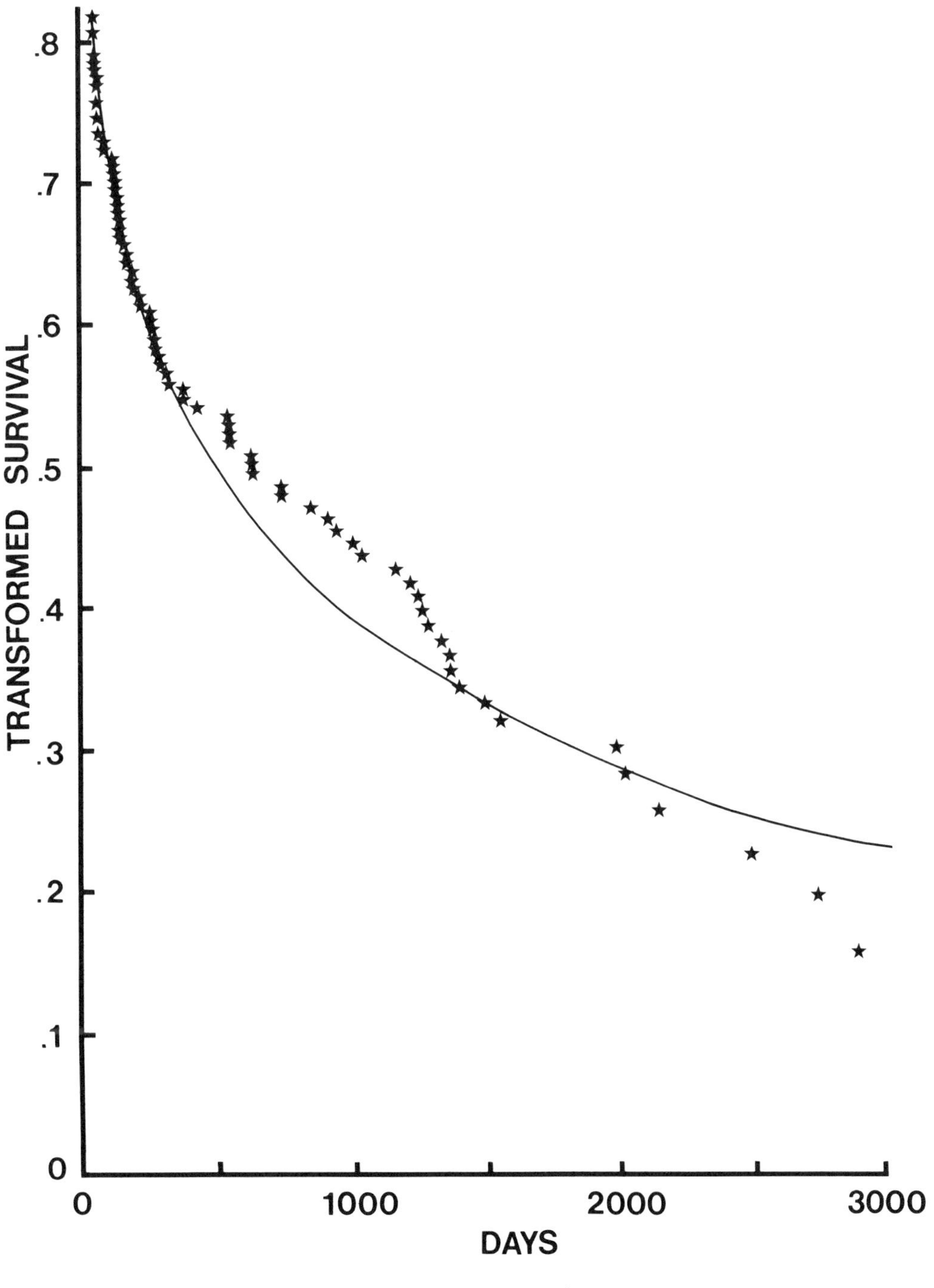

FIGURE 2: Estimated survival functions $\hat{S}$ from (6) and the Kaplan-Meier estimate.

4. Large Sample Properties

For ease of exposition, we establish our results for trials conducted over a fixed time period [0,T] and where items enter only at one of the K fixed times

$$0 = t_1 < t_2 < \cdots < t_K < T .$$

Let n_i = number of items entering at time t_i and $n = \sum_{i=1}^{K} n_i$. Setting $L_i = T - t_i$ and $\eta_i = \ell im(n_i/n)$, we introduce

$$\begin{aligned} H_{\underset{\sim}{\eta}}(\lambda,\mu,\sigma) = \sum_{i=1}^{K} \eta_i \Bigg\{ & [1 - G(L_i)] \, \ell n \left[\frac{1 - \Phi\left(\frac{L_i^{(\lambda)} - \mu}{\sigma}\right)}{1 - G(L_i)} \right] \\ & + E_g\left(I_{[0,L_i]} \, \ell n \left[\frac{\phi\left(\frac{X_{i1}^{(\lambda)} - \mu}{\sigma}\right) X_{i1}^{\lambda - 1}}{g(X_{i1})} \right] \right) \Bigg\} , \end{aligned} \tag{7}$$

which is the negative of the Kullback-Leibler information number between $X^{(\lambda)}$ and some normal distribution obtained by weighting the K censored population numbers by the proportions η_i.

THEOREM 1:

Let $n_i/n \to \eta_i$ $(0 < \eta_i < 1)$ and suppose the following conditions are satisfied:

(i) the parameter space Ω is the compact subset of R^3 defined by $\Omega = \{\theta = (\lambda,\mu,\sigma)' \mid |\mu| \le M, \ c_1 \le \sigma \le c_2, \ a \le \lambda \le b$ for some $0 < M, c_1, c_2, b < \infty$ and $-\infty < a < 0\}$,

(ii) the moments $E_g(X^{2a})$ and $E_g(X^{2b})$ are finite,

(iii) $H_{\underset{\sim}{\eta}}(\lambda,\mu,\sigma)$ has a unique global maximum at $(\lambda_0,\mu_0,\sigma_0) = \underset{\sim}{\theta}_0'$.

Then, (1) $(\hat{\lambda},\hat{\mu},\hat{\sigma}) \xrightarrow{a.s.} (\lambda_0,\mu_0,\sigma_0)$ as $n \to \infty$.

Furthermore, if:

(iv) $(\lambda_0,\mu_0,\sigma_0)$ is an interior point of Ω,

(v) both $E_g[X^a \; \ell n(X)]^2$ and $E_g[X^b \; \ell n(X)]^2$ are finite,

(vi) $\nabla H_{\underset{\sim}{\eta}}(\lambda_0,\mu_0,\sigma_0) = \underset{\sim}{0}$,

(vii) $V = \{\nabla^2 H_{\underset{\sim}{\eta}}(\lambda_0,\mu_0,\sigma_0)\}^{-1}$ exists,

then, (2) $\sqrt{n}\,(\hat{\lambda}-\lambda_0,\; \hat{\mu}-\mu_0,\; \hat{\sigma}-\sigma_0)' \xrightarrow{d} N_3(\underset{\sim}{0},VWV')$ as $n \to \infty$, with the elements of $W = (w_{uv})$, $u,v = 1,2,3$, given by

$$\sum_{i=1}^{K} \eta_i \left\{ [1-G(L_i)] \left(\frac{\partial}{\partial\theta_u} \ell n \left[1-\Phi\left(\frac{L_i^{(\lambda)} - \mu}{\sigma} \right) \right] \Bigg|_{\underset{\sim}{\theta}_0} \right) \left(\frac{\partial}{\partial\theta_v} \ell n \left[1-\Phi\left(\frac{L_i^{(\lambda)} - \mu}{\sigma} \right) \right] \Bigg|_{\underset{\sim}{\theta}_0} \right) \right.$$

$$\left. + E_g \left[I_{[0,L_i]}(X) \left(\frac{\partial \; \ell n \; \phi\left(\frac{X^{(\lambda)} - \mu}{\sigma} \right) X^{\lambda-1}}{\partial\theta_u} \Bigg|_{\underset{\sim}{\theta}_0} \right) \left(\frac{\partial \; \ell n \; \phi\left(\frac{X^{(\lambda)} - \mu}{\sigma} \right) X^{\lambda-1}}{\partial\theta_v} \Bigg|_{\underset{\sim}{\theta}_0} \right) \right] \right\} .$$

PROOF: The log-likelihood, divided by the sample size $n = \sum_{i=1}^{K} n_i$, is the sum of the K terms

$$n^{-1} \ell_n = \sum_{i=1}^{K} \frac{n_i}{n} \left\{ \sum_{j=1}^{n_i} \left[-\frac{1}{2}\,\ell n(2\pi) - \frac{1}{2}\,\ell n\,\sigma - \frac{1}{2}\left(\frac{x_{ij}^{(\lambda)} - \mu}{\sigma} \right)^2 + (\lambda-1)\ell n(x_{ij}) \right] \right.$$

$$\left. \cdot\; I_{[0,L_i]}(x_{ij}) + m_i \; \ell n \left[1-\Phi\left(\frac{L_i^{(\lambda)} - \mu}{\sigma} \right) \right] \right\}$$

where $m_i = \sum_{j=1}^{n_i} I_{(L_i,\infty]}(x_{ij})$. The term $\left(\frac{x^{(\lambda)}}{\sigma} - \mu\right)^2 I_{[0,L_i]}(x)$

(i) is dominated by a g-integrable function uniformly in $\underset{\sim}{\theta} = (\lambda,\mu,\sigma)' \in \Omega$ and

(ii) is equicontinuous in $\underset{\sim}{\theta}$ for fixed x, on the set $S_m = [0, L_i - \frac{1}{m}] \cup [L_i + \frac{1}{m}, m]$.

The uniform strong law (see Rubin (1956)) applies.

$$\frac{1}{n_i} \sum_{j=1}^{n_i} \left(\frac{X_{ij}^{(\lambda)} - \mu}{\sigma}\right)^2 I_{[0,L_i]}(X_{ij}) \xrightarrow{\text{a.s.}} E_g \left[\left(\frac{X_{i1}^{(\lambda)} - \mu}{\sigma}\right)^2 I_{[0,L_i]}(X_{i1})\right]$$

for $i = 1,2,\ldots,K$ uniformly in (λ,μ,σ). The strong law of large numbers applied to $\sum_{j=1}^{n_i} \ell n(X_{ij})/n_i$ and $\sum_{j=1}^{n_i} I_{(L_i,\infty]}(X_{ij})/n_i$, for $i = 1,2,\ldots,K$, establishes the almost sure uniform convergence

$$\frac{1}{n} \ell_n \to H_{\underset{\sim}{\eta}}(\lambda,\mu,\sigma) + \sum_{i=1}^{K} \eta_i \{E_g[I_{[0,L_i]} \ell n g(X_{1i})] + [1 - G(L_i)] \ell_n [1-G(L_i)]\} \ .$$

Moreover, $H_{\underset{\sim}{\eta}}(\lambda,\mu,\sigma)$ is continuous and, by assumption, has a unique maximum at $(\lambda_0,\mu_0,\sigma_0)$. Consequently, $(\hat{\lambda},\hat{\mu},\hat{\sigma}) \to (\lambda_0,\mu_0,\sigma_0)$ almost surely.

The asymptotic normality follows upon expanding the first partial derivatives of $n^{-\frac{1}{2}}\ell_n$ in a Taylor series about $(\lambda_0,\mu_0,\sigma_0)$. Since $(\hat{\lambda},\hat{\mu},\hat{\sigma}) \to (\lambda_0, \mu_0,\sigma_0)$, which is interior to Ω, $\nabla\ell_n(\hat{\lambda},\hat{\mu},\hat{\sigma}) = \underset{\sim}{0}$ for all sufficiently large n. Similar to the treatment of the single sample problem in Guerrero (1979) and Guerrero and Johnson (1979), we can dominate the individual terms in $n^{-1}\nabla^2\ell_n(\lambda,\mu,\sigma)$ to obtain uniform convergence to its expected value $\nabla^2 H_{\underset{\sim}{\eta}}$. In particular, $n^{-1}\nabla^2 \ell n(\lambda_*,\mu_*,\sigma_*)$ converges a.s. to $\nabla^2 H_{\underset{\sim}{\eta}}(\lambda_0,\mu_0,\sigma_0)$ where $(\lambda_*,\mu_*,\sigma_*)$ is any sequence of intermediate values between $(\hat{\lambda},\hat{\mu},\hat{\sigma})$ and $(\lambda_0,\mu_0,\sigma_0)$. Since

$$n^{-\frac{1}{2}} \nabla\ell_n(\lambda_0,\mu_0,\sigma_0) \xrightarrow{\mathcal{L}} N(\underset{\sim}{0},W)$$

the normal convergence for $(\hat{\lambda},\hat{\mu},\hat{\sigma})'$ follows.

5. Checking the Adequacy of the Transformation

The power transformation was selected by maximizing the likelihood (5) obtained under the tentative assumption that some transformation (1) is normal. Although Theorem 1 gives one set of conditions that insure that, asymptotically, the power transformation closest to a normal is selected, this choice may not be good enough. Therefore, it is necessary to check that the transformation

$$x^{(\hat{\lambda})} = \frac{x^{\hat{\lambda}} - 1}{\hat{\lambda}}$$

has achieved near-normality.

In order to obtain diagnostic plots, a censored observation $x_i = L_i$ is assigned the value

$$\hat{x}_i^{(\hat{\lambda})} = E[X^{(\hat{\lambda})} | X > L_i] = \hat{\mu} + \hat{\sigma}\, h\left(\frac{L_i^{(\hat{\lambda})} - \hat{\mu}}{\hat{\sigma}}\right) , \tag{8}$$

where $h(\cdot) = \phi(\cdot) / [1 - \Phi(\cdot)]$ is the hazard rate for the standard normal (see Schmee and Hahn (1979)). That is, the expectation is computed as if $X^{(\hat{\lambda})}$ is normal with mean $\hat{\mu}$ and variance $\hat{\sigma}^2$. Using these estimates, we have

$$\begin{aligned} &x_i^{(\hat{\lambda})} && \text{, if failure} \\ &\hat{x}_i^{(\hat{\lambda})} = \hat{\mu} + \hat{\sigma}\, h\left(\frac{L_i^{(\hat{\lambda})} - \hat{\mu}}{\hat{\sigma}}\right) && \text{, if censored.} \end{aligned}$$

These can be ordered and displayed in a normal Q-Q plot.

Figure 3 shows a plot of the transformed heart transplant survival times versus the approximate normal scores $\Phi^{-1}(i/(n+1))$. The predicted values, plotted as open squares, do not seem to conform to the straight line pattern. At this state of development, it is not clear that (8) provides the proper estimates for diagnostic plots. Figure 4 shows the transformed failure times by themselves, plotted against the same scores as in Figure 3. It is these uncensored observations on which the adequacy of the normal approximation should be judged.

In the complete sample situation, goodness-of-fit can be tested using the correlation coefficient calculated from the normal Q-Q plot. Verrill (1981) has recently determined the large sample distribution of the correlation coefficient calculated from data that are right-censored. His results apply to either fixed order statistic censoring or fixed time censoring but not to the staggered entry situation graphed in Figure 3.

6. Survival Analysis Setting with Covariates

When r predictors $\underset{\sim}{z}' = (z_1,\ldots,z_r)$ are available, the tentative assumption becomes

$$X^{(\lambda)} \text{ is distributed } N(\alpha + \underset{\sim}{\beta}'\underset{\sim}{z}, \sigma^2)$$

for some choice of λ. Under this assumption, the likelihood becomes

$$L(\lambda,\alpha,\underset{\sim}{\beta},\sigma) = \prod_{i\varepsilon F} \frac{1}{(2\pi)^{\frac{1}{2}}\sigma} e^{-\frac{1}{2\sigma^2}(x_i^{(\lambda)} - \alpha - \underset{\sim}{\beta}'\underset{\sim}{z}_i)^2} x_i^{\lambda-1}$$

$$\cdot \prod_{i \notin F} \left[1 - \Phi\left(\frac{\frac{L_i^{\lambda}-1}{\lambda} - \alpha - \underset{\sim}{\beta}'\underset{\sim}{z}_i}{\sigma}\right)\right] \quad . \tag{9}$$

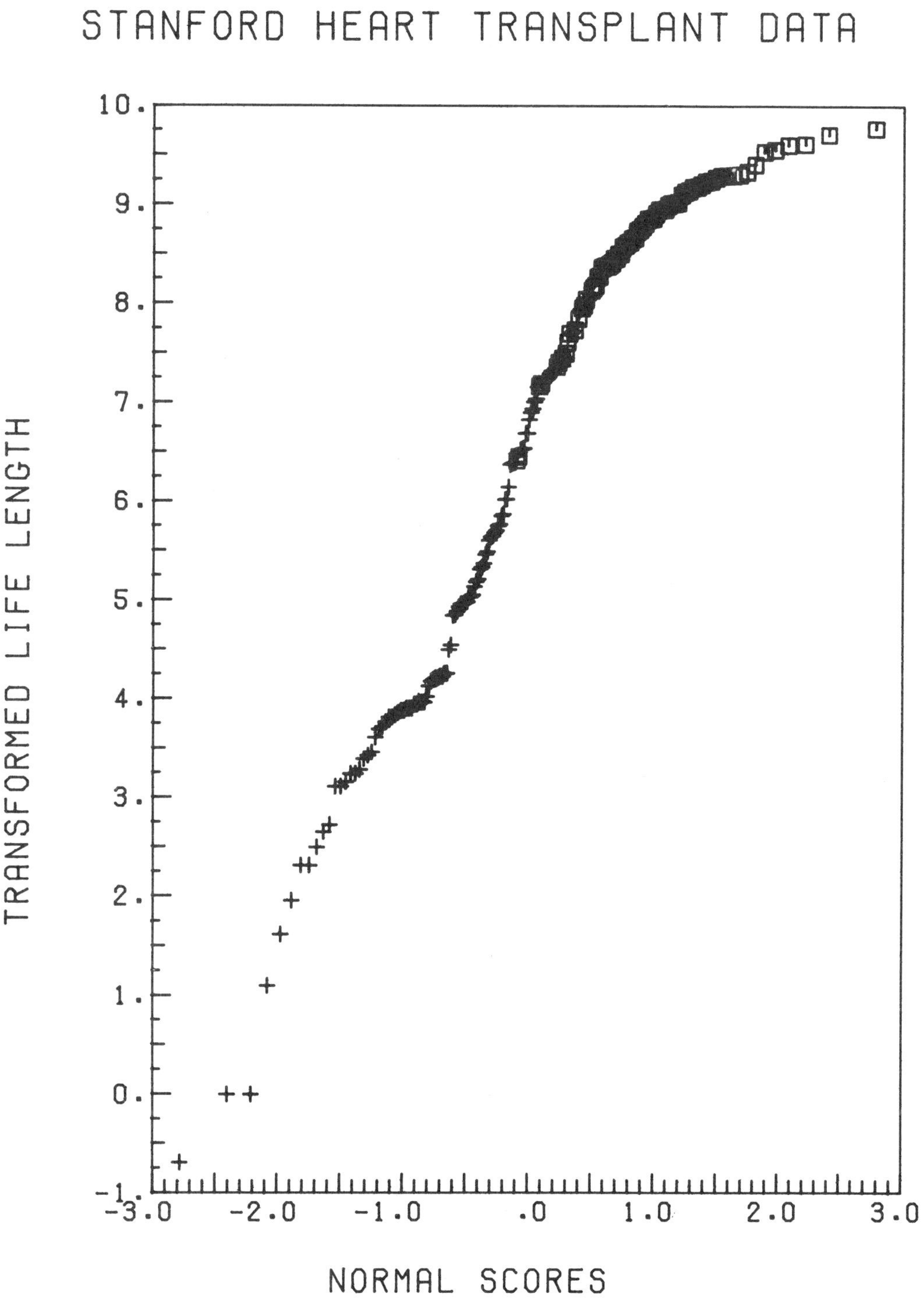

FIGURE 3: Normal scores plot of heart transplant data.

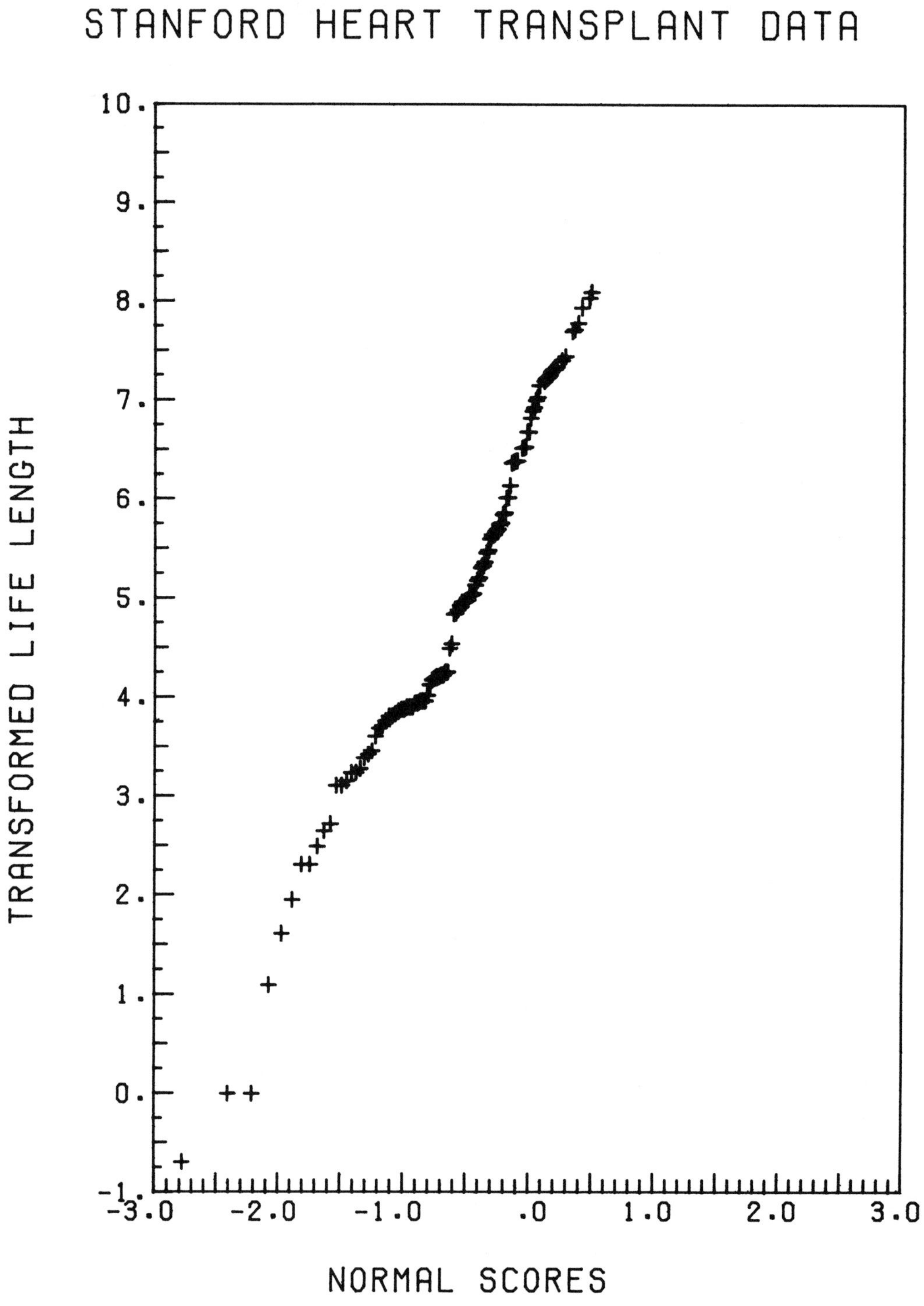

FIGURE 4: Portion of normal scores plot from death times.

Our parametric approach is to maximize (9). Miller (1980) contains a discussion of several alternative methods for formulating the regression model.

EXAMPLE:

We return to the Stanford heart-transplant data and use $z = \text{age}$ as a predictor variable. A computer calculation provides the estimates

$$\hat{\lambda} = .0090\ , \quad \hat{\alpha} = 7.9339\ , \quad \hat{\beta} = -.0349\ , \quad \hat{\sigma} = 2.5490$$

and

$$-\ell n\ L(\hat{\lambda},\hat{\alpha},\hat{\beta},\hat{\sigma}) = 857.3343\ .$$

The maximized likelihood for $\lambda = 0$, the log-transformation, is nearly the same. Note also that $-2\ \ell n[L(\hat{\lambda},\hat{\mu},\hat{\sigma}) / L(\hat{\lambda},\hat{\alpha},\hat{\beta},\hat{\sigma})]$ is less than $\chi_1^2\ (.05)$, suggesting that age is not a good predictor.

Regression diagnostics need to be developed to check both the normal assumption for $X^{(\lambda)}$ and the regression equation. To obtain plots, we replace each censored value by its conditional expected value

$$\hat{x}^{(\hat{\lambda})} = \hat{\alpha} + \hat{\underset{\sim}{\beta}}'\underset{\sim}{z} + \hat{\sigma}\, h\left(\frac{x^{(\hat{\lambda})} - \hat{\alpha} - \hat{\underset{\sim}{\beta}}'\underset{\sim}{z}}{\hat{\sigma}}\right)\ . \tag{10}$$

The residuals, divided by $\hat{\sigma}$, are then

$$\hat{\varepsilon} = \begin{cases} \dfrac{x^{(\hat{\lambda})} - \hat{\alpha} - \hat{\underset{\sim}{\beta}}'\underset{\sim}{z}}{\hat{\sigma}}\ , & \text{if failure} \\[2ex] h\left(\dfrac{x^{(\hat{\lambda})} - \hat{\alpha} - \hat{\underset{\sim}{\beta}}'\underset{\sim}{z}}{\hat{\sigma}}\right)\ , & \text{if censored}\ . \end{cases} \tag{11}$$

A normal Q-Q plot of $\hat{\varepsilon}$ for the heart transplant data looks almost identical to Figure 3. A plot of $\hat{\varepsilon}$ versus $x^{(\hat{\lambda})}$ (or $\hat{x}^{(\hat{\lambda})}$) is shown as Figure 5 where squares represent the residuals from censored observations. A plot of $\hat{\varepsilon}$ versus patient number is given as Figure 6. Note how the predicted residuals from the later cases in the study form a bounding curve.

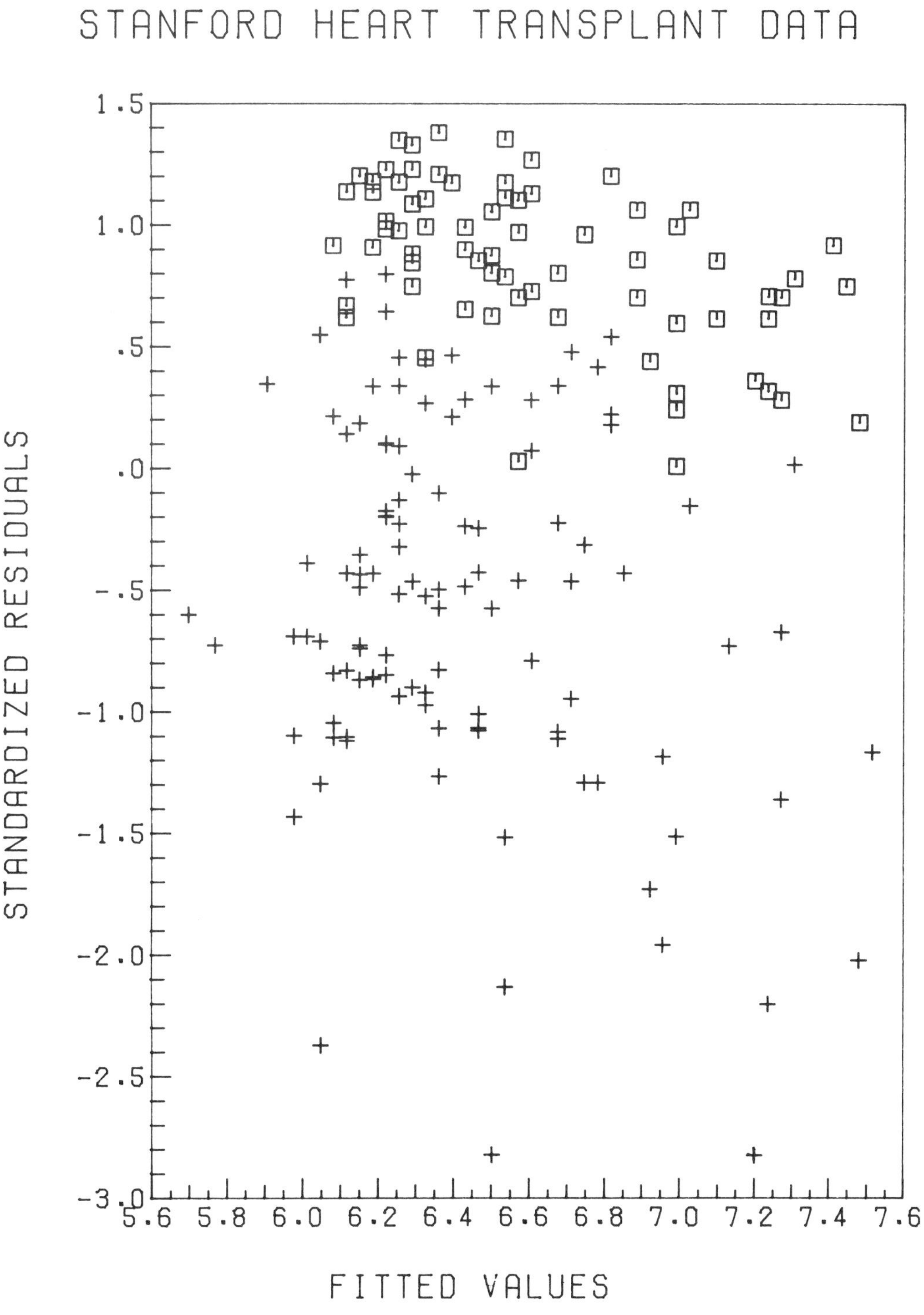

FIGURE 5: Standardized residuals versus fitted values when age is a covariate. ⊡ Censored value.

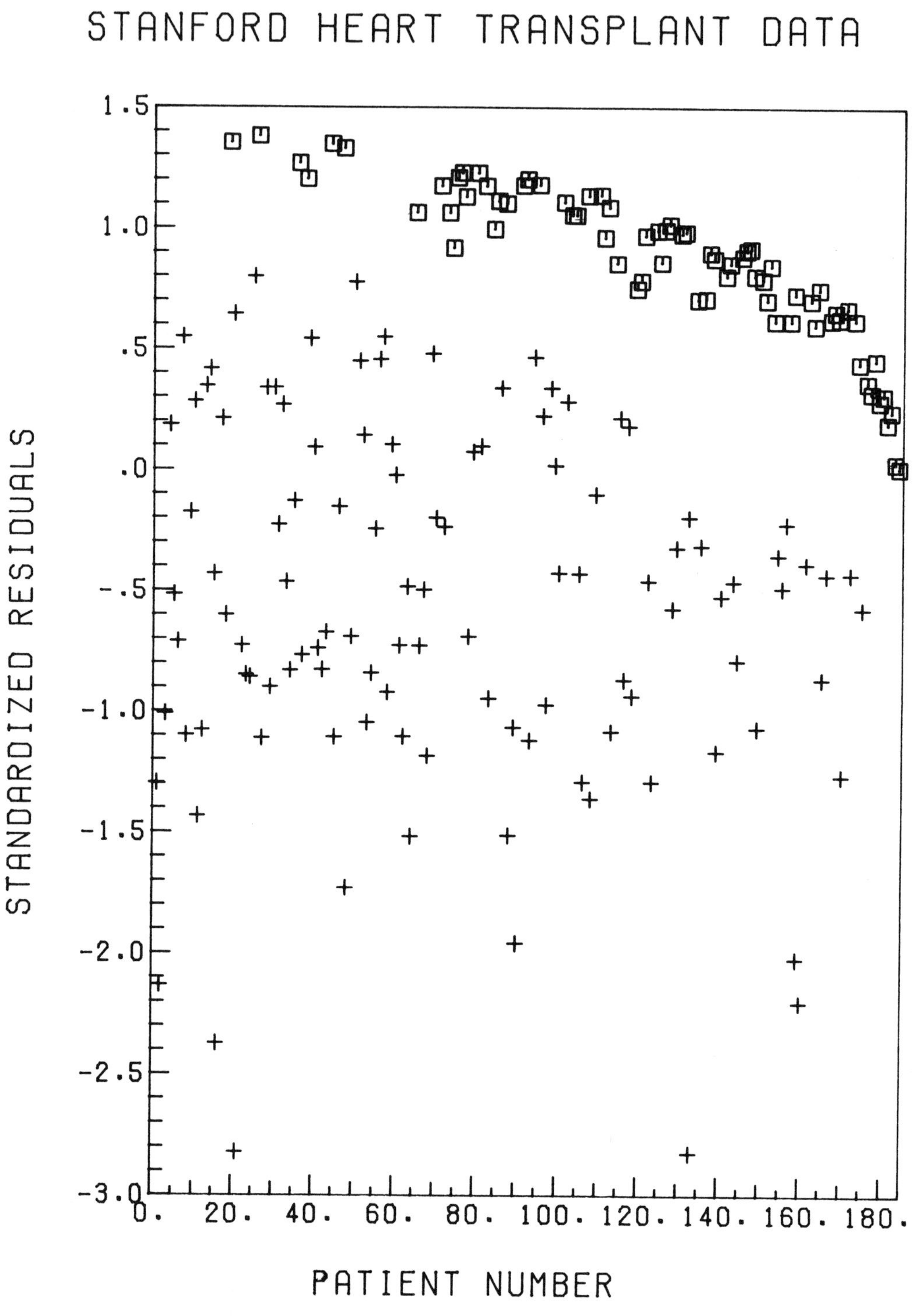

FIGURE 6: Standardized residuals versus patient number.
⊡ Censored value.

ACKNOWLEDGEMENT

This research was sponsored by the Office of Naval Research under Grant No. N00014-78-C-0722.

REFERENCES

Bickel, P.J. and Doksum, K. (1981). An analysis of transformations revisited. Journal of the American Statistical Association, **76**, 296-311.

Box, G.E.P. and Cox, D.R. (1964). An analysis of transformations. Journal of the Royal Statistical Society B, **26**, 211-243.

Buckley, J. and James, I. (1979). Linear regression with censored data. Biometrika **66**, 429-436.

Carroll, R.J. (1980). A robust method for testing transformations to achieve approximate normality. Journal of the Royal Statistical Society B, **42**, 71-78.

Carroll, R.J. and Ruppert, D. (1981). On prediction and the power transformation family. Biometrika **68**, 609-615.

Cox, D.R. (1972). Regression models and life tables (with discussion). Journal of the Royal Statistical Society B, **34**, 187-202.

Guerrero, V. (1979). Extensions of the Box-Cox transformation to grouped-data situations. Ph.D. Thesis, Department of Statistics, University of Wisconsin.

Guerrero, V. and Johnson, R.A. (1979). Transformation of grouped or censored data to near normality. Technical Report 542, Department of Statistics, University of Wisconsin-Madison.

Hernandez, F. and Johnson, R.A. (1980). The large-sample behavior of transformations to normality. Journal of the American Statistical Association, **75**, 855-861.

Koul, H., Susarla, V. and Van Ryzin, J. (1981). Regression analysis with randomly right censored data. Annals of Statistics **9**, 1276-1288.

Miller, R.G. (1981). Survival analysis. John Wiley, New York.

Miller, R.G. (1976). Least squares regression with censored data. Biometrika **63**, 449-464.

Miller, R.G. and Halpern, J. (1981). Regression with censored data. Technical Report No. 66, Division of Biostatistics, Stanford University.

Rubin, H. (1956). Uniform convergence of random functions with applications to statistics. Annals of Mathematical Statistics **27**, 200-203.

Schmee, J. and Hahn, G.J. (1979). A simple method for regression analysis with censored data. Technometrics **21**, 417-432.

COVARIATE MEASUREMENT ERRORS IN THE ANALYSIS OF COHORT AND CASE-CONTROL STUDIES

Ross Prentice

Fred Hutchinson Cancer Research Center and Department of Biostatistics, University of Washington

1. Introduction

This paper discusses the analysis of 'failure' time data, when predictor variables are subject to measurement error. The author's symposium presentation concentrated on a partial likelihood approach to relative risk estimation when covariates are subject to measurement error; material that mostly will appear in Prentice (1982). To avoid undue repetition the presentation here will emphasize full likelihood and marginal likelihood approaches to this problem. The accommodation of covariate measurement errors in the context of case-control sampling will also be briefly considered.

In failure time studies, as well as in many other areas of application, covariate values are subject to measurement errors. Particular applications that motivated this work include a study of the relationship between radiation exposure level and cancer mortality in atomic bomb survivors and a study of cardiovascular disease risk factors in a large cohort study. In the former study, one is interested in cancer mortality dose-response effects corresponding to individual gamma and neutron exposures. These exposure level estimates were, however, imputed from distance (from the presumed hypocenter) and shielding information obtained by interview. Such estimates may differ substantially from the 'true' exposure levels; in fact, the quality of the dosimetry data has recently been the subject of much controversy (e.g., Marshall, 1981). In the cardiovascular disease study, data on covariates such as blood pressure, serum cholesterol level and leukocyte counts were obtained in biennial clinic visits, taking place over a 20-year period. Each of these measured predictor variables is subject to considerable variation partially due to limitations of the measuring process but primarily because a large number of additional factors influence the measured values. For example, one may be interested in the relationship between some intrinsic blood pressure level and coronary heart disease incidence, but the measured blood pressure may be a rather imprecise approximation thereto, since it depends so heavily on the person's recent activities, state of relaxation, and current position in the diurnal cycle, to name a few factors. In many studies it will be possible to make some reasonable specifications of the error distributions associated with covariate measurements. Whether or not there is much basis for such specification, the sensitivity of results to various error distribution assumptions will be of interest.

2. Induced Models and Parametric Estimation

Consider a failure time random variable $T \geq 0$ and, for the moment, a fixed covariate $z = (z_1, \ldots, z_p)$. Throughout f will be used generically to denote probability, or probability density, function, so that $f(t|z)$ denotes the conditional density for T given z. Characteristics, such as relative risk parameters, used in the specification of $f(t|z)$ will usually be the primary target of estimation. Now suppose that, rather than z, one observes only the 'measured' covariate $x = (x_1, \ldots, x_q)$. Usually there will be a one-to-one correspondence between components of x and z, and $p = q$, but this is not required in the discussion that follows. In considering error distribution assumptions it is natural to think of a specification of the distribution of x given z, along with a marginal distribution for z. As will be seen below, however, it is only necessary to specify the conditional probability distribution, $f(z|x)$, for z given x, rather than their joint distribution, in order to proceed with estimation of $f(t|z)$.

We will require a conditional independence between T and x, given z; that is,

$$f(t|z,x) = f(t|z) \quad . \tag{1}$$

This condition is a statement that the measured covariate has no prognostic value if the true covariate is known. If (1) does not hold, x is not simply an 'estimator' of z and direct modelling of $f(t|z,x)$ is indicated.

The induced probability function for T given the measured covariate x is readily derived as the expectation over the distribution of z given x of $f(t|z,x)$, which under (1) can be written

$$f(t|x) = E_x\{f(t|z)\} \quad . \tag{2}$$

If the error distribution $f(z|x)$ is completely specified, this induced model $f(t|x)$ will involve only the parameters of $f(t|z)$. It is then of interest to identify failure time and error distribution models that lead to tractable induced models for failure time, given the observable covariate. Such induced models can then be applied to failure time data in order to carry out inferences on parameters of interest.

In order to develop mathematically convenient induced models (2), it is natural to consider normally distributed failure time and error random variables. Suppose $Y = \log T$ satisfies the normal linear regression model

$$Y = \log T = \alpha + z\beta + \sigma V \quad ,$$

where α, $\sigma > 0$ and $\beta(p\times 1)$ are real parameters and V is a standard normal random variable. Also, suppose that the true covariate distribution, given the corresponding measurement x, is normal with mean vector (1×p) μ_x and variance matrix $\sum_x$. The induced model (2) for $Y = \log T$ given x is then readily shown to be normal with mean vector $\alpha + \mu_x\beta$ and variance $\sigma^2 + \beta' \sum_x \beta$, where β' denotes the vector transpose of β. This simple result may provide an adequate basis for exploring the implications of covariate measurement errors on regression testing and estimation, in a variety of failure time and non-failure time applications. Specifically, an iterative maximum likelihood procedure could be readily implemented for β estimation, that would not be unduly complicated by the presence of right censorship.

Other distributional assumptions may also yield explicit induced models. For example, a Weibull regression model with 'linear' hazard ratio $(1+z\beta) \geq 0$, can be written

$$\lambda(t|z) = \lambda p(\lambda t)^{p-1} (1+z\beta) \quad ,$$

where λ denotes the hazard, or instantaneous failure rate function, and λ, $\sigma > 0$ and β(p×1) are parameters. A normal distribution for z given x yields, after some algebra, an induced hazard function

$$\lambda(t|x) = \lambda p(\lambda t)^{p-1} [1 + \{\mu_x - (\lambda t)^p \beta' \textstyle\sum_x\} \beta] \quad .$$

In fact, some bounds on the support for z given x will be required in order that $1+z\beta \geq 0$ not be violated. A normal model for z given x, and the above induced model $\lambda(t|x)$ should, however, provide adequate approximations if β, μ_x and $\sum_x$ are such that $1+z\beta \geq 0$ with probability close to one at each x. Note that the hazard ratio corresponding to any pair of x-values is no longer constant, but rather converges monotonically to unity as $t \to \infty$. Computational methods for fitting this induced model could be derived, though the model is perhaps too complicated to expect much use.

The fitting of such models to failure time data will, as usual, require an independent censorship assumption in order that (2) be identifiable. Such an assumption can be written

$$\lambda\{t|x, \text{no censorship in } [0,t)\} = \lambda(t|x) \quad . \tag{3}$$

In some problems an independent censorship assumption applied to t given z, rather than t given x, would be more appropriate. In such circumstances censoring will typically be mildly dependent and (2) will not strictly be identifiable. This seems unlikely to be a practical problem, however, unless covariate errors are very substantial and censorship depends heavily on z.

In order to use standard likelihood expression one will also require the independence of failure times given the corresponding measured covariate x_i, $i = 1,\ldots,n$. Such independence will follow, for example, if z_i, $i = 1,\ldots,n$ can be viewed as i.i.d. from some distribution and both t_i given z_i and x_i given z_i are independent for $i = 1,\ldots,n$.

It seems appropriate to make some comment on the specification of the probability function $f(z|x)$. For example, in order to specify the mean μ_x and variance matrix Σ_x in the above normal densities, one might suppose that the basic regression vector z can be viewed as normally distributed with mean μ and variance Σ and that the measured covariate x arises via $x = z+w$, where w is normal with mean zero and variance matrix C. If z and w are independent the density for z given x is then normal with mean $\mu_x = \mu + \Sigma(\Sigma + C)^{-1}(x-\mu)$ and variance matrix $\Sigma_x = \Sigma - \Sigma(\Sigma + C)^{-1}\Sigma$, the latter of which is independent of x. In the normal regression model described above (for $Y = \log T$) the induced regression equation in x will then have regression coefficient $\Sigma(\Sigma + C)^{-1}\beta$. Ignoring covariate measurement, errors would then give rise to coefficient estimates that are systematically too close to zero in simple linear regression and that are 'deflated' by the matrix $\Sigma(\Sigma + C)^{-1}$ in the multiple regression problem. More generally, z and w may be allowed to be correlated. A normal distribution for z given x is readily derived from any joint normal distribution for z and w. In the very special case, sometimes referred to as the Berkson model, z and w have a joint normal distribution as above except that the covariance of z and w is -C. It follows that $\mu_x = x$ and $\sum_x = C$. It is worth noting again that specification of the joint distribution of z and x is unnecessary, since only the distribution of z given x appears in (2). Joint normal distributions, of the type just described, may then be used as a guide toward the specification of μ_x and $\sum_x$ in a normal model for z given x, but do not need to be explicitly assumed. The reader is referred to the review paper, Cochran (1968), for further comments on error distribution specification and on the effects of measurement errors in the ordinary regression model.

3. Cox Model Estimation with Covariate Errors

The models described above, particularly the induced log-normal model for T given x, provide the basis for a parametric approach to accommodating measurement errors in failure time analyses. The partially parametric regression model of Cox (1972) is an attractive alternative to failure time analyses. Desirable features include the ability to interpret the regression parameter in terms of relative risk, substantial model flexibility, and the availability of many important generalizations, as summarized in Kalbfleisch and Prentice (1980). In its most general form the method gives a computationally feasible method of exploring the dependence of the relative risk function on covariates and follow-up time, without placing any model restrictions, except the presumed parametric form for the relative risk function.

The special case of the Cox model in which the relative risk is independent of t can be written

$$\lambda(t|z) = \lambda_0(t)\ g(z\beta) \quad , \tag{4}$$

where $\lambda_0(\cdot) \geq 0$ is unrestricted, $g(\cdot) \geq 0$ is a specified function standardized so that $g(0) = 1$ and $\beta(p\times 1)$ is a regression parameter to be estimated. Note that $g(z\beta) = \lambda(t|z)/\lambda(t|z=0)$ is the risk associated with regression vector z, relative to that at $z = 0$. Usually the relative risk function has been defined by $g(u) = \exp(u)$, though other forms such as $g(u) = 1 + u$ have also sometimes been used.

Various approaches have been considered for the estimation of β in (4); most notably, the partial likelihood approach of Cox (1972,1975). Kalbfleisch and Prentice (1973) utilized a marginal likelihood approach that was based on the distribution of failure time ranks.

In the presence of covariate errors the model induced from (4) via (2) has the rather complicated corresponding hazard function

$$(5) \qquad \lambda(t|x) = \lambda_0(t)\left[\int g(z\beta)\exp\{-g(z\beta)\int_0^t \lambda_0(u)du\} f(z|x)dx \Big/ \int \exp\{-g(z\beta)\int_0^t \lambda_0(u)du\} f(z|x)dz\right],$$

where the integrals (or sums) are over the range of z, given x. In the special case $g(z\beta) = 1+z\beta$, and z given x normal with mean μ_x and variance matrix $\sum_x$, (5) simplifies to

$$\lambda(t|x) = \lambda_0(t)[1 + \{\mu_x - \int_0^t \lambda_0(u)du\ \beta' \textstyle\sum_x\}\beta] \quad ,$$

generalizing the Weibull regression result given above. In spite of the complexity of (5), the induced class of models retains the property of functional invariance under monotone-increasing differentiable transformations on t. One can show, as in Kalbfleisch and Prentice (1973), that the distribution of the failure time ranks does not involve the baseline hazard function $\lambda_0(\cdot)$. In fact, the failure time rank vector is marginally sufficient for β in the sense described by Kalbfleisch and Prentice. Assuming the expectation operators and order statistic integrals in the generalized rank vector probability can be interchanged, the marginal likelihood for β in (5) can be written

$$L(\beta|X) = E_X L(\beta|Z) \quad , \tag{6}$$

where X has been written for the set of measured covariate vectors over the sample; that is, $X = \{x_1,\ldots,x_n\}$, Z has been written for $\{z_1,\ldots,z_n\}$, the expectation is over the distribution of $Z = \{z_1,\ldots,z_n\}$ given $X = \{x_1,\ldots,x_n\}$ and $L(\beta|Z)$ is the marginal likelihood that would arise if the true covariate vectors Z, rather than only X, were observed. Specifically,

$$L(\beta|Z) = \prod_{i=1}^{k} \left[\prod_{\ell \varepsilon F(t_i)} g(z_\ell \beta) \Big/ \left\{ \sum_{\ell \varepsilon R(t_i)} g(z_\ell \beta) \right\}^{m_i} \right] \quad , \tag{7}$$

where $t_1,\ldots,t_k$ represent the distinct (uncensored) failure times in the sample, $F(t_i)$ is the set of $m_i \geq 1$ study subjects that fail at t_i and $R(t_i)$ is the risk set just prior to time t_i. Note that the denominator of (7) involves an approximation (Breslow, 1974) to accommodate any tied failure times. The score statistic from (6) is

$$\partial \log L(\beta|X)/\partial\beta = L(\beta|X)^{-1} E_X\{L(\beta|Z)\ \partial \log L(\beta|Z)/\partial\beta\} \quad , \tag{8}$$

a weighted average of the score statistics corresponding to possible values of Z, given X. Similarly, the observed information matrix can be written

$$-\partial^2 \log L(\beta|X)/\partial\beta^2 = \tag{9}$$

$$L(\beta|X)^{-1} E_X[L(\beta|Z)\ \{-\partial^2 \log L(\beta|Z)/\partial\beta^2 - \partial \log L(\beta|Z)/\partial\beta'\ \partial \log L(\beta|Z)/\partial\beta]$$

$$+ \{\partial \log L(\beta|X)/\partial\beta'\}\{\partial \log L(\beta|X)/\partial\beta\} \quad .$$

At $\beta = 0$, $L(\beta|Z)$ is independent of Z and (8) simplifies to

$$(10) \qquad g'(0) \sum_{i=1}^{k} \left\{ \sum_{\ell \in F(t_i)} E_{x_\ell}(z_\ell) - m_i n_i^{-1} \sum_{\ell \in R(t_i)} E_{x_\ell}(z_\ell) \right\} ,$$

where n_i is the number of subjects in $R(t_i)$. It follows that it is only necessary to specify the expectations of each z-value given the corresponding x-value in order to carry out a score test for $\beta = 0$, a point that was made somewhat more generally, in the context of partial likelihood, in Prentice (1982).

In order to use (6) for general inference on the regression parameter β one needs to contend with a complicated expectation. The possibility of developing useful analytic expressions for (6) seems remote, even if mathematically convenient choices for g and $f(z|x)$ are entertained. An approximate estimation procedure, based on Monte Carlo sampling is suggested by (8) and (9). In particular, suppose that sets of regression vectors $Z_1, \ldots, Z_s$ are sampled from the joint distributions of Z given X. The score statistic (8) is then estimated by

$$\tilde{v} = \sum_{j=1}^{s} L(\beta|Z_j) \, \partial \log L(\beta|Z_j)/\partial\beta \Big/ \sum_{j=1}^{s} L(\beta|Z_j)$$

while the corresponding observed information matrix is estimated by

$$\sum_{j=1}^{s} L(\beta|Z_j) \, \{-\partial^2 \log L(\beta|Z_j) - \partial \log L(\beta|Z_j)/\partial\beta' \, \partial \log L(\beta|Z_j)/\partial\beta\} \Big/ \sum_{j=1}^{s} L(\beta|Z_j) + \tilde{v}' \, \tilde{v} \quad .$$

Existing computer software could then be readily adapted to carry out a Newton-Raphson maximization for β. This idea amounts simply to approximating (6) by

$$s^{-1} \sum_{j=1}^{s} L(\beta | Z_j) \quad .$$

As such, the approximation can be made to be as close as desired by making s large. This is a computation-intensive approach to regression estimation. It is, however, very flexible in terms of both the relative risk function, g , and error distribution $f(z|x)$. In fact, it is not even necessary that z-values on distinct study subjects be independent, given the corresponding measured x-values. It is hoped to pursue this idea in more detail elsewhere.

A nonparametric maximum likelihood approach to estimation in (4) would lead in the presence of measurement errors to a likelihood function for β that can again be written

$$\tilde{L}(\beta | X) = E_X \tilde{L}(\beta | Z) \quad ,$$

where $\tilde{L}(\beta|Z)$ is the likelihood function, given the 'true' covariate values Z, after maximizing out the baseline hazard function. The approximate likelihood of Breslow (1974) would lead once again to (6), in the presence of covariate errors.

Prentice (1982) considered a partial likelihood approach to this problem. A partial likelihood function for β, given the measured covariate values, X , in the sample can be written

$$(11) \qquad \prod_{i=1}^{k} \left[\prod_{\ell \varepsilon F(t_i)} E_{(t_i, x_\ell)} \, g(z_\ell \beta) \Big/ \left\{ \sum_{\ell \varepsilon R(t_i)} E_{(t_i, x_\ell)} \, g(z_\ell \beta) \right\}^{m_i} \right] \quad ,$$

where a tied failure time approximation has again been made and the expectations in the ith term of the product are conditional on both $T \geq t_i$ and the measured covariate values. The partial likelihood approach accommodates time-dependent covariates as may be defined to test or relax the proportional hazards assumption in (4) or may be utilized to relate failure rate to some stochastic covariate process. The result (10) could equally well be derived from (11). Unfortunately, however, (11) does not provide an adequate answer to more general testing and estimation problems in many applications, since the expectations in (11) typically involve the baseline incidence function $\lambda_0(\cdot)$, as is evident from (5) upon noting that

$$\lambda(t|x) = \lambda_0(t)\, E_{(t,x)}\, g(z\beta) \quad .$$

The application that motivated Prentice (1982) was such that the dependence of $E_{(t,x)}\, g(z\beta)$ on the condition $T \geq t$, and hence the dependence of the expectation on $\lambda_0(\cdot)$, could be ignored. If such dependence cannot be ignored, it would be useful to consider iterative estimation procedures in which a trial value of β is used to produce an empirical estimate of the cumulative hazard function $\Lambda_0(t) = \int_0^t \lambda_0(u)du$ that appears in (5), which in turn, is used to obtain an updated β-value on the basis of (11). Even in the simple special case $g(u) = 1+u$ with normally distributed covariate errors, a nonparametric maximum likelihood approach to estimating $\Lambda_0(t)$, at a specified β, is complicated. On the other hand, unless covariate errors are quite substantial, it would presumably be accurate enough to obtain an empirical estimate of $\Lambda_0(t)$, ignoring covariate measurement errors, and subsequently use this esimate in the partial likelihood function (11). Such a usage would be quite routine, for example, in the circumstances mentioned above in which z, given x, is normally distributed and the relative risk function g is of a linear form. Numerical evaluation of this proposal would be worthwhile.

4. Covariate Errors in Case-Control Studies

Suppose now that a Cox-type model (4) holds for the incidence (or mortality) rate for a disease. A case-control study involves selecting both diseased (cases) and disease-free (control) subjects and sampling their corresponding covariate data z. Often z will include summarizations of certain exposure histories along with personal characteristics. In the presence of covariate measurement errors one will sample the measured covariate x, rather than z.

To be specific, consider the type of case-control study described in Prentice and Breslow (1978), in which for each case a set of time (age) matched controls are selected. The hazard function induced from (4) can, in general, be written

$$\lambda(t|x) = \lambda_0(t)\, E_{(t,x)}\, g(z\beta) \quad ; \tag{12}$$

that is, the induced relative risk of time t is the expectation of $g(z\beta)$, given the measured covariate x and given $T \geq t$. By the same argument used in Prentice and Breslow, a conditional likelihood for this relative risk function can be developed by conditioning on the set of exposure histories corresponding to each case and its matched controls. The conditional likelihood function can be written

$$L(\beta) = \prod_{i=1}^{k} E_{(t_i,x_i)}\, g(z_i\beta) \Big/ \sum_{\ell \in R_i} E_{(t_i,x_\ell)}\, g(z_\ell\beta) \quad , \tag{13}$$

where $t_1,\ldots,t_k$ denote the incidence times for the cases and R_i denotes the ith case and its matched controls. As with the partial likelihood described previously, however, the induced relative risk function will depend to some extent on the baseline incidence function $\lambda_0(\cdot)$ due to the inclusion of $\{T \geq t_i\}$ in the conditioning event. If, however, the study disease is rare the distribution of z-values that correspond to a measured covariate x will be very similar among subjects without failure at some time t as was the case at $t = 0$. In this circumstance, the relative risk function will be well approximated by

(14) $$E_x\, g(z\beta)$$

and straightforward asymptotic likelihood procedures can be applied to (13) for β estimation. In the special case $g(u) = 1+u$ one then fits a relative risk model

(15) $$1 + E_x(z|x)$$

to the case-control data using (13). As a practical approach to accommodating covariate measurement errors in the estimation of exposure-response relationships Armstrong and Oakes (1982) have suggested replacing z-values by corresponding $E_x(z|x)$ values, and carrying out standard analyses. With a linear relative risk function, their proposal is supported by the development given here provided the condition $\{T \geq t\}$ can be ignored in the induced relative risk function. Estimation with a multiplicative relative risk function $g(u) = \exp(u)$ can be readily carried out with error probability functions $f(z|x)$ that have simple moment generating functions. For example, the normal probability function for z given x mentioned above, gives for (14)

$$\exp\{\mu_x\beta + \tfrac{1}{2}\beta' \textstyle\sum_x \beta\} \quad ,$$

which may be inserted into (13) for estimation of β.

Similar results could be developed for more general case-control study designs and, for example, logistic disease incidence models.

5. Concluding Remarks

Failure to acknowledge covariate measurement errors in some regression problems may lead to results that lack a useful interpretation or that are misleading. Greater effort seems warranted in respect to methods to estimate covariate error distribution properties and to utilize information on covariate error distributions toward the estimation of key regression parameters. This paper described several approaches to the latter problem in a failure time regression context. Clearly the surface has merely been scratched on this important statistical topic.

ACKNOWLEDGEMENTS

This work was supported by grants GM-28314 and GM-24472 from the National Institutes of Health.

REFERENCES

Armstrong, B.G. and Oakes, D. (1982). Effects of approximation in exposure assessments on estimates of exposure-response relationships. To appear, Scandinavian Journal of Work, Environment and Health.

Breslow, N.E. (1974). Covariance analysis of censored survival data. Biometrics **30**, 89-99.

Cochran, W.G. (1968). Errors of measurement in statistics. Technometrics **10**, 637-666.

Cox, D.R. (1972). Regression models and life tables (with discussion). Journal of the Royal Statistical Society B, **34**, 187-220.

Cox, D.R. (1975). Partial likelihood. Biometrika **62**, 269-276.

Kalbfleisch, J.D. and Prentice, R.L. (1973). Marginal likelihoods based on Cox's regression and life model. Biometrika **60**, 267-278.

Kalbfleisch, J.D. and Prentice, R.L. (1980). The Statistical Analysis of Failure Time Data. New York, Wiley.

Marshall, E. (1981). New A-bomb studies alter radiation estimates. Science **212**, 900-903.

Prentice, R.L. (1982). Covariate measurement errors and parameter estimation in Cox's failure time regression model. To appear, Biometrika.

Prentice, R.L. and Breslow, N.E. (1978). Retrospective studies and failure time models. Biometrika 65, 153-158.

CONFIDENCE BOUNDS FOR THE EXPONENTIAL MEAN IN TIME-TRUNCATED LIFE TESTS

N.R. Mann
University of California, Los Angeles

R.E. Schafer
Hughes Aircraft Company

M.C. Han
University of California, Los Angeles

1. Introduction

A commonly occurring life-test situation is: a time T is specified, n units are put on test without replacement and the successive ordered times-to-failure $X_1 \leq \cdots \leq X_r < T$, $r \leq n$, are observed. This life testing procedure is commonly referred to as Type 1 censoring, which will be assumed throughout this paper.

Here we suppose that each of the n units tested has the same one-parameter exponential life-time distribution of which the mean is θ. Computing methods will be developed for the lower confidence bound on θ based on the maximum-likelihood estimate (MLE).

The MLE, say $\hat{\theta}$, has been given by Halperin (1950), Bartlett (1953a,b), Deemer and Votaw (1955) and Bartholomew (1957):

(1) $$\hat{\theta} = [\sum_{i=1}^{r} X_i + (n-r)\,T\,]\,/\,r, \quad r \geq 1 \quad .$$

When $r = 0$, $\hat{\theta}$ is undefined.

Approximate confidence intervals have been studied by Bartlett (1953a,b) and Bartholomew (1963), and the asymptotic properties of $\hat{\theta}$ have been investigated by Deemer and Votaw (1955) and Yang and Sirvanci (1977). In particular, $\hat{\theta}$ is consistent and asymptotically unbiased, i.e., $\lim_{n\to\infty} E_c(\hat{\theta}) = \theta$, where E_c denotes the conditional expectation on $r > 0$.

The exact distribution of $\hat{\theta}$ has had an interesting history: Halperin (1950) gave the distribution of $\hat{\theta}$, conditional on r, and described very briefly how the unconditional distribution could be obtained; Halperin (1960) gave the distribution of $r\hat{\theta}$; Bartholomew (1963) was the first to give the distribution of $\hat{\theta}(r \geq 1)$; Hoem (1969) essentially presented the distribution of $\hat{\theta}$ again, along with other results.

Barlow, et al. (1968) developed a computer program for obtaining interval estimates of θ, and Spurrier and Wei (1980) presented a hypothesis test procedure based on $\hat{\theta}$, in which $r = 0$ is not conditioned out. In each of these two papers the exact distribution of $\hat{\theta}$ was used. Virtually all authors have commented on the computational complexity of the exact distribution. For example, it was seventeen years (1963 to 1980) from the availability of the exact distribution of $\hat{\theta}$ to the development of hypothesis tests for this most important life-testing situation.

The inclusion of $r = 0$ in the hypothesis test procedure of Spurrier and Wei (1980) is tantamount to taking $\hat{\theta} = \infty$ (i.e., always accept $H_o: \theta \geq \theta_o$) when $r = 0$. In that sense it should be noted that the two distribution functions are simply related:

$$(2) \qquad P(\hat{\theta} \geq t \mid r \geq 1) = \frac{P(\hat{\theta} \geq t \mid r \geq 0) - e^{-(n\theta^{-1}T)}}{1 - e^{-(n\theta^{-1}T)}} .$$

As indicated by the numerator in (2) the inclusion of $r = 0$ in testing $H_o: \theta \leq \xi$ $(\theta \geq \xi)$ precludes test sizes $\alpha \leq \exp(-n\xi^{-1}T)$ $((1-\alpha) < \exp(-n\xi^{-1}T))$. On the other hand, inclusion of $r = 0$ means that the test is always applicable.

Since $\hat{\theta}$ is undefined when $r = 0$, in this case one can use the fact that the number of failures r is a binomial random variable with parameters n and $p = 1 - \exp(-T/\theta)$ to obtain a confidence bound. Hence, for confidence level $1 - \alpha$,

$$\theta_*/T = n(-\ln\alpha)^{-1} \quad ,$$

is a $(1 - \alpha)$-level lower confidence bound for θ/T.

The case of a random sample of size n from an exponential distribution (right) truncated at T is different from that considered here. The former case has been considered in some detail by, among others, Bain, et al. (1977) and Deemer and Votaw (1955).

2. Exact Confidence Bounds for θ

Hypothesis tests involve a critical region. Hence the results of Spurrier and Wei (1980) cannot be used for obtaining interval estimates which require direct use of the distribution function as in the computer program of Barlow, et al. (1968).

Following Barlow, et al., if the confidence coefficient is $1 - \alpha$, then based on Bartholomew's (1963) result θ_* must satisfy

$$1 - \alpha = (1 - e^{-nT/\theta_*})^{-1} \sum_{k=1}^{n} \binom{n}{k} e^{-(n-k)T/\theta_*} \sum_{i=0}^{k} \binom{k}{i} (-1)^i e^{-iT/\theta_*}$$

(3)

$$\times\ \phi(2[k\hat{\theta} - (n - k + i)T\,]/\theta_*,\ 2k) \quad ,$$

where

$$\phi(u,\nu) = \begin{cases} [2^{\nu/2}\,\Gamma(\nu/2)]^{-1} \int_0^u e^{-t/2}\, t^{(\nu/2)-1}\, dt & u \geq 0 \\ 0, & u < 0 \end{cases} ,$$

and $r \geq 1$.

Inspecting (3), one notes that five quantities must be specified to obtain θ_*: $n, T, r, \hat{\theta}$ and $1-\alpha$ and thus it appears that a computer program must be used for each different estimating situation. However, the following points are noted:

i) the random variable r is needed solely to compute $\hat{\theta}$.

ii) the variables T and θ_* appear always as the ratio $(\theta_*/T)^{-1}$.

iii) $[k\hat{\theta} - (n-k+i)T]/\theta_* = [\frac{k\hat{\theta}}{T} - (n-k+i)](T/\theta_*)$
and hence $\hat{\theta}$ appears only with T as $\hat{\theta}/T$ and, again,
θ_* and T appear only as the ratio $(\theta_*/T)^{-1}$.

iv) only a few confidence coefficients (i.e., 0.90, 0.95) are commonly used.

v) n, $\hat{\theta}$ and T are known.

Thus, $\hat{\theta}/T$ can be an entry variable (with n and $1-\alpha$) to obtain θ_*/T. Multiplication by T then yields θ_*.

3. Computational Aspects and an Approximation

A computer program along the lines given by Barlow, et al. (1968) was written for the purpose of tabulating values of θ_*/T. It produced results agreeing exactly with their results (to four significant digits) for the two examples given by those authors, namely $\theta_* = 28.49$ and $\theta_* = 32.09$, for $\hat{\theta} = 51.166$, $T = 50$, $n = 10$ and $\alpha = 0.10$ and 0.05, respectively. Results of the computer program also agreed rather well with the asymptotic normal and chi-square approximations in an example given by Bartholomew (1963) for $\hat{\theta}/T = 0.705$, $n = 20$, $\alpha = 0.025$. Bartholomew's asymptotic lower confidence bounds on θ/T are 0.45 and 0.46. The computer program yields 0.39.

It was found, however, that in another example given by Bartholomew (1963) for $\hat{\theta}/T = 3.35$, $n = 40$, $\alpha = 0.025$, the asymptotic lower bound is 1.9, while the computer program yields 0.63 for θ_*/T, the "exact" lower confidence bound.

The vast discrepancy in these latter results appears to be due to a combination of factors. First the chi-square subroutine used in the computer program produces results that are accurate to about 10^{-3} for $\hat{\theta}/T$ less than 1. As $\hat{\theta}/T$ approaches n (where $\hat{\theta}/T < n$), the accuracy in the output of the subroutine declines. Second, $(n+1)(n+2)/2$ terms involving these chi-square evaluations are summed. Thus, when $n = 40$, there are 861 such terms to be summed. Another possible explanation, of course, is that a sample size of 40 is too small for asymptotic results to apply.

To attempt to ascertain the level of accuracy of the estimate provided by the computer program, three avenues are explored. A more presice chi-square evaluation, from IBM, with 10^{-9} accuracy was incorporated in the program. In addition, a method was found for calculating approximate confidence bounds which do not depend on asymptotic values. This allowed for comparisons to be made for small values of n. Finally a simulation study was performed.

We noted above the asymptotic chi-square approximation used by Bartholomew. This is simply a two-moment fit that uses the conditional (on $r > 0$) mean m and variance v of $\hat{\theta}$ (see Patnaik (1949)): $2m\hat{\theta}/v$ is approximately a chi-square variate with $\nu = 2m^2/v$ degrees of freedom. Bartholomew used the asymptotic conditional mean and variance, respectively, of $\hat{\theta}$ for m and v and noted that for T infinite, $2m\hat{\theta}/v$ is an exact chi-square variate with 2n degrees of freedom since, in this case, the expectation and variance of $\hat{\theta}$ are θ and θ^2/n, respectively.

Yang and Sirvanci (1977) have provided expressions for the exact conditional mean and variance of $\hat{\theta}$. These can readily be converted to exact conditional moments of $\hat{\theta}/T$, involving only the parameter θ/T. Dividing the expressions for the conditional mean and variance of $\hat{\theta}$ given by those authors by T and T^2, respectively, we obtain

$$E_c(\hat{\theta}/T) = \theta/T - 1/p + nE_c(1/r)$$

and

$$\mathrm{Var}_c(\hat{\theta}/T) = E_c(1/r)(\theta^2/T^2 - q/p^2) + n^2\,\mathrm{Var}_c(1/r) \ ,$$

where

$$q = e^{-T/\theta}, \ p = 1 - q \ ,$$

$$E_c(1/r) = \sum_{k=1}^{n} \binom{n}{k} p^k q^{n-k}/[k(1-q^n)] \ ,$$

and

$$E_c(1/r^2) = \sum_{k=1}^{n} \binom{n}{k} p^k q^{n-k}/[k^2(1-q^n)] \ .$$

In these expressions $E_c(\cdot)$ and $\mathrm{Var}_c(\cdot)$ indicate conditioning on $r > 0$. In the sequel we let $m = E_c(\hat{\theta}/T)$ and $v = \mathrm{Var}_c(\hat{\theta}/T)$.

Use of the chi-square approximation results generally in noninteger degrees of freedom. Thus, to obtain iteratively the appoximate lower confidence bound for θ/T we use the Wilson-Hilferty (1931) transformation of chi-square to normality, namely, for $\nu = 2m^2/v$,

$$3\sqrt{\nu/2}\ [\left(\frac{\hat{\theta}/T}{m}\right)^{1/3} + 2/(9\nu) - 1\ ]$$

is approximately N(0,1). This transformation has been used in several instances by Mann, Schafer and Singpurwalla (1974), particularly in obtaining approximate lower confidence bounds for scale parameters. McGinnis and Sammons (1970), in an investigation of gamma approximations, showed that the Wilson-Hilferty and Severo and Zelen (1960) equations are most effective for our purposes. Their results and results of Mann, Schafer and Singpurwalla indicate that the Wilson-Hilferty approximation yields acceptable results (at least two good significant figures) as long as the number of degrees of freedom $\nu = 2m^2/v$ is 3.5 or greater.

A comparison was made of the exact lower confidence bounds on θ/T (with confidence level $1 - \alpha$) obtained from the computer program with the IBM chi-square subroutine and the bounds based on the chi-square approximation, for

$n = 5(5)20$, $\hat{\theta}/T = n/2$, $n/4$, 1, $1/n$, $2/n$, $4/n$ and $\alpha = 0.10$, 0.05 and 0.01. This resulted in discrepancies of 2 or less in the second to fourth significant figures in the corresponding values of θ_*/T for $n = 10$, 15, 20 and $\hat{\theta}/T$ less than or equal to 1, and very large discrepancies for large values of $\hat{\theta}/T$. In all of the cases evaluated, $\nu = 2m^2/v > 3.5$.

A simulation study was then undertaken to determine the accuracy of the values of θ_*/T computed by the two methods. In the study we evaluated the probability p associated with obtaining values less than $\hat{\theta}/T$ when θ_*/T is the true value of θ/T. When θ_*/T (associated with confidence level $1 - \alpha$) is the correct value, p is equal to α.

A Monte Carlo sample size of 5000 was used with various combinations of specified values of n, $\hat{\theta}/T$ and α, including those chosen for the earlier comparison of the two methods. Particular attention was paid to very large and very small values of $\hat{\theta}/T$.

The specified values for the confidence levels are essentially correct (i.e., deviations in $\hat{p}$, the calculated p, from the specified value of α are within expected bounds) for the "exact" values of θ_*/T calculated by means of the computer program for $\alpha = 0.1$, 0.05 and 0.01 when $n = 2(1)10$ and $\hat{\theta}/T < n - 0.2$ and when $n = 15$, 20 and $\hat{\theta}/T \leq 1$. Thus, it appears that an increase in sample size beyond 10 causes problems for the computer-program estimator of θ_*/T unless $\hat{\theta}/T$ is quite small. Tabulations for $n = 2(1)10$ appear in Table 1.

The asymptotic estimator of θ_*/T also works best for small values of $\hat{\theta}/T$, as indicated earlier by the comparisons made with the computer-program estimates. A simulation made to examine the asymptotic bounds for $n = 40$ gave results that confirmed this conclusion. For $\hat{\theta}/T = 0.1$, 0.2, 0.5, 1(1)5, $\hat{p}$ is within expected bounds for $\alpha = 0.1$ and 0.05. For $\alpha = 0.01$ and for values of $\hat{\theta}/T$ 10 or larger the values obtained for θ_*/T are not accurate.

TABLE 1. Values of θ_*/T, lower confidence bound at level $1-\alpha$, corresponding to $\hat{\theta}/T$, for $n=2(1)10$, $\alpha=0.1$, 0.05, 0.01.

n	$\hat{\theta}/T$	α=0.1	α=0.05	α=0.01
2	1.8	1.4575	0.7913	0.4127
2	1.7	0.9256	0.6120	0.3580
2	1.6	0.7259	0.5180	0.3217
2	1.5	0.6118	0.4552	0.2935
2	1.4	0.5330	0.4075	0.2694
2	1.3	0.4724	0.3681	0.2476
2	1.2	0.4224	0.3336	0.2271
2	1.1	0.3790	0.3023	0.2076
2	1.0	0.3397	0.2729	0.1885
2	0.9	0.3388	0.2725	0.1883
2	0.8	0.3345	0.2701	0.1876
2	0.7	0.3230	0.2625	0.1842
2	0.6	0.2991	0.2445	0.1740
2	0.5	0.2572	0.2108	0.1507
2	0.4	0.2056	0.1686	0.1206
2	0.3	0.1542	0.1264	0.0904
2	0.2	0.1028	0.0843	0.0603
2	0.1	0.0514	0.0421	0.0301
3	2.8	2.2646	1.2154	0.6347
3	2.7	1.4559	0.9543	0.5579
3	2.6	1.1540	0.8209	0.5092
3	2.5	0.9885	0.7341	0.4728
3	2.4	0.8780	0.6703	0.4436
3	2.3	0.7959	0.6196	0.4181
3	2.2	0.7304	0.5770	0.3952
3	2.1	0.6757	0.5399	0.3741
3	2.0	0.6282	0.5066	0.3542
3	1.5	0.6289	0.5062	0.3540
3	1.4	0.6254	0.5043	0.3526
3	1.3	0.6118	0.4955	0.3478
3	1.2	0.5846	0.4761	0.3367
3	1.1	0.5418	0.4439	0.3167
3	1.0	0.4844	0.3991	0.2871
3	0.9	0.4273	0.3541	0.2566
3	0.8	0.3798	0.3161	0.2301
3	0.7	0.3383	0.2823	0.2059
3	0.6	0.3001	0.2512	0.1836
3	0.5	0.2655	0.2235	0.1651
3	0.4	0.2240	0.1894	0.1418
3	0.3	0.1690	0.1429	0.1071
3	0.2	0.1127	0.0953	0.0714
3	0.1	0.0563	0.0476	0.0357
4	3.8	3.0718	1.6410	0.8567
4	3.7	1.9890	1.2958	0.7574
4	3.6	1.5826	1.1227	0.6959
4	3.5	1.3641	1.0117	0.6513
4	3.4	1.2213	0.9312	0.6156
4	3.3	1.1166	0.8684	0.5855
4	3.2	1.0346	0.8166	0.5593
4	3.1	0.9672	0.7724	0.5355
4	3.0	0.9096	0.7333	0.5136
4	2.0	0.9091	0.7338	0.5124
4	1.9	0.9026	0.7289	0.5103
4	1.8	0.8800	0.7119	0.5017
4	1.7	0.8385	0.6811	0.4842
4	1.6	0.7789	0.6365	0.4563
4	1.5	0.7060	0.5812	0.4207
4	1.4	0.6395	0.5303	0.3874
4	1.3	0.5895	0.4912	0.3609
4	1.2	0.5510	0.4606	0.3391
4	1.1	0.5219	0.4372	0.3218
4	1.0	0.5011	0.4212	0.3107
4	0.9	0.4749	0.4015	0.2989
4	0.8	0.4318	0.3667	0.2753
4	0.7	0.3792	0.3228	0.2437
4	0.6	0.3282	0.2803	0.2123
4	0.5	0.2809	0.2406	0.1830
4	0.4	0.2329	0.2002	0.1534
4	0.3	0.1794	0.1546	0.1193
4	0.2	0.1197	0.1032	0.0797
4	0.1	0.0599	0.0516	0.0398
5	4.8	3.7948	2.0679	1.0784
5	4.7	2.5209	1.6372	0.9567
5	4.6	2.0113	1.4243	0.8823
5	4.5	1.7396	1.2888	0.8290
5	4.4	1.5641	1.1915	0.7869
5	4.3	1.4366	1.1161	0.7521
5	4.2	1.3376	1.0547	0.7220
5	4.1	1.2570	1.0029	0.6952
5	4.0	1.1889	0.9577	0.6708
5	2.5	1.1889	0.9567	0.6701
5	2.4	1.1807	0.9489	0.6666
5	2.3	1.1489	0.9250	0.6532
5	2.2	1.0925	0.8839	0.6288
5	2.1	1.0158	0.8275	0.5938
5	2.0	0.9269	0.7618	0.5524
5	1.9	0.8493	0.7041	0.5155
5	1.8	0.7933	0.6619	0.4876
5	1.7	0.7529	0.6309	0.4661
5	1.6	0.7262	0.6098	0.4507
5	1.5	0.7138	0.6003	0.4437
5	1.4	0.7056	0.5942	0.4408
5	1.3	0.6826	0.5759	0.4303
5	1.2	0.6389	0.5410	0.4071
5	1.1	0.5832	0.4958	0.3750
5	1.0	0.5284	0.4510	0.3423
5	0.9	0.4841	0.4144	0.3152
5	0.8	0.4446	0.3820	0.2917
5	0.7	0.4017	0.3468	0.2668
5	0.6	0.3508	0.3037	0.2354
5	0.5	0.2964	0.2573	0.2004
5	0.4	0.2429	0.2115	0.1656
5	0.3	0.1865	0.1629	0.1283
5	0.2	0.1251	0.1093	0.0862
5	0.1	0.0625	0.0546	0.0431
6	5.8	4.4991	2.4950	1.2999
6	5.7	3.0492	1.9783	1.1558
6	5.6	2.4395	1.7256	1.0685

TABLE 1 (continued)

n	$\hat{\theta}/T$	α=0.1	α=0.05	α=0.01	n	$\hat{\theta}/T$	α=0.1	α=0.05	α=0.01
6	5.5	2.1148	1.5656	1.0064	7	2.3	1.1101	0.9301	0.6935
6	5.4	1.9065	1.4513	0.9578	7	2.2	1.1105	0.9285	0.6932
6	5.3	1.7562	1.3634	0.9181	7	2.1	1.0944	0.9230	0.6873
6	5.2	1.6401	1.2924	0.8840	7	2.0	1.0589	0.9005	0.6710
6	5.1	1.5461	1.2327	0.8539	7	1.9	1.0011	0.8541	0.6397
6	5.0	1.4671	1.1810	0.8268	7	1.8	0.9310	0.7957	0.6008
6	3.0	1.4730	1.1776	0.8266	7	1.7	0.8660	0.7416	0.5641
6	2.9	1.4610	1.1676	0.8213	7	1.6	0.8205	0.7026	0.5372
6	2.8	1.4168	1.1380	0.8032	7	1.5	0.7925	0.6780	0.5198
6	2.7	1.3442	1.0867	0.7723	7	1.4	0.7690	0.6607	0.5071
6	2.6	1.2508	1.0187	0.7303	7	1.3	0.7338	0.6342	0.4881
6	2.5	1.1465	0.9421	0.6822	7	1.2	0.6813	0.5905	0.4572
6	2.4	1.0575	0.8770	0.6416	7	1.1	0.6259	0.5432	0.4231
6	2.3	0.9949	0.8308	0.6118	7	1.0	0.5783	0.5031	0.3937
6	2.2	0.9510	0.7980	0.5905	7	0.9	0.5347	0.4662	0.3668
6	2.1	0.9228	0.7767	0.5754	7	0.8	0.4842	0.4243	0.3355
6	2.0	0.9116	0.7682	0.5694	7	0.7	0.4295	0.3774	0.2997
6	1.9	0.9121	0.7671	0.5696	7	0.6	0.3751	0.3308	0.2641
6	1.8	0.9088	0.7625	0.5682	7	0.5	0.3188	0.2821	0.2265
6	1.7	0.8913	0.7501	0.5601	7	0.4	0.2592	0.2299	0.1856
6	1.6	0.8495	0.7190	0.5389	7	0.3	0.1976	0.1757	0.1425
6	1.5	0.7897	0.6716	0.5061	7	0.2	0.1329	0.1182	0.0961
6	1.4	0.7266	0.6205	0.4702	7	0.1	0.0665	0.0591	0.0480
6	1.3	0.6744	0.5776	0.4396	8	7.8	6.9127	3.3347	1.7392
6	1.2	0.6378	0.5471	0.4176	8	7.7	4.0841	2.6551	1.5504
6	1.1	0.6068	0.5213	0.3998	8	7.6	3.2888	2.3237	1.4382
6	1.0	0.5686	0.4901	0.3782	8	7.5	2.8615	2.1167	1.3594
6	0.9	0.5169	0.4477	0.3472	8	7.4	2.5879	1.9689	1.2982
6	0.8	0.4637	0.4027	0.3134	8	7.3	2.3930	1.8555	1.2488
6	0.7	0.4148	0.3614	0.2828	8	7.2	2.2427	1.7653	1.2064
6	0.6	0.3644	0.3188	0.2511	8	7.1	2.1222	1.6902	1.1696
6	0.5	0.3093	0.2713	0.2151	8	7.0	2.0211	1.6254	1.1368
6	0.4	0.2517	0.2215	0.1763	8	4.0	2.0256	1.6210	1.1375
6	0.3	0.1925	0.1698	0.1358	8	3.9	1.9982	1.6135	1.1283
6	0.2	0.1293	0.1142	0.0915	8	3.8	1.9327	1.5703	1.1002
6	0.1	0.0647	0.0571	0.0458	8	3.7	1.8346	1.4963	1.0558
7	6.8	5.7191	2.9202	1.5206	8	3.6	1.7136	1.4030	1.0007
7	6.7	3.5715	2.3176	1.3542	8	3.5	1.5811	1.3034	0.9420
7	6.6	2.8672	2.0259	1.2535	8	3.4	1.4704	1.2218	0.8935
7	6.5	2.4891	1.8419	1.1835	8	3.3	1.3946	1.1664	0.8593
7	6.4	2.2480	1.7106	1.1285	8	3.2	1.3431	1.1285	0.8349
7	6.3	2.0752	1.6102	1.0837	8	3.1	1.3108	1.1047	0.8191
7	6.2	1.9420	1.5294	1.0456	8	3.0	1.2971	1.0963	0.8145
7	6.1	1.8345	1.4618	1.0121	8	2.6	1.3091	1.0924	0.8161
7	6.0	1.7446	1.4036	0.9822	8	2.5	1.2898	1.0941	0.8121
7	3.5	1.7565	1.3979	0.9823	8	2.4	1.2602	1.0817	0.8009
7	3.4	1.7326	1.3885	0.9754	8	2.3	1.2138	1.0407	0.7751
7	3.3	1.6764	1.3532	0.9523	8	2.2	1.1440	0.9785	0.7357
7	3.2	1.5905	1.2911	0.9140	8	2.1	1.0686	0.9137	0.6934
7	3.1	1.4828	1.2104	0.8653	8	2.0	1.0087	0.8606	0.6580
7	3.0	1.3642	1.1227	0.8123	8	1.9	0.9703	0.8262	0.6350
7	2.9	1.2644	1.0495	0.7678	8	1.8	0.9466	0.8100	0.6217
7	2.8	1.1953	0.9988	0.7358	8	1.7	0.9247	0.8012	0.6128
7	2.7	1.1476	0.9637	0.7127	8	1.6	0.8986	0.7776	0.5978
7	2.6	1.1172	0.9413	0.6984	8	1.5	0.8511	0.7330	0.5683
7	2.5	1.1050	0.9331	0.6926	8	1.4	0.7896	0.6819	0.5320

TABLE 1 (continued)

n	$\hat{\theta}/T$	α=0.1	α=0.05	α=0.01	n	$\hat{\theta}/T$	α=0.1	α=0.05	α=0.01
8	1.3	0.7368	0.6395	0.5002	9	0.7	0.4525	0.4027	0.3270
8	1.2	0.6960	0.6071	0.4758	9	0.6	0.3941	0.3519	0.2870
8	1.1	0.6539	0.5725	0.4505	9	0.5	0.3336	0.2988	0.2450
8	1.0	0.6022	0.5276	0.4183	9	0.4	0.2711	0.2434	0.2004
8	0.9	0.5479	0.4817	0.3835	9	0.3	0.2060	0.1853	0.1532
8	0.8	0.4970	0.4379	0.3503	9	0.2	0.1384	0.1247	0.1034
8	0.7	0.4424	0.3917	0.3146	9	0.1	0.0693	0.0624	0.0517
8	0.6	0.3852	0.3420	0.2761	10	9.8	7.2291	4.0400	2.1375
8	0.5	0.3267	0.2910	0.2363	10	9.7	5.0197	3.2750	1.9199
8	0.4	0.2656	0.2372	0.1936	10	9.6	4.0725	2.8907	1.7921
8	0.3	0.2021	0.1808	0.1482	10	9.5	3.5632	2.6405	1.6932
8	0.2	0.1358	0.1217	0.1000	10	9.4	3.2425	2.4694	1.6269
8	0.1	0.0680	0.0609	0.0500	10	9.3	3.0072	2.3331	1.5681
9	8.8	8.0543	3.7199	1.9450	10	9.2	2.8279	2.2286	1.5205
9	8.7	4.5771	2.9826	1.7424	10	9.1	2.6885	2.1405	1.4782
9	8.6	3.7006	2.6184	1.6211	10	9.0	2.5645	2.0611	1.4396
9	8.5	3.2250	2.3868	1.5324	10	5.0	2.5532	2.0766	1.4407
9	8.4	2.9212	2.2236	1.4653	10	4.9	2.5196	2.0631	1.4275
9	8.3	2.7061	2.0996	1.4111	10	4.8	2.4393	1.9999	1.3911
9	8.2	2.5400	1.9987	1.3653	10	4.7	2.3179	1.9025	1.3342
9	8.1	2.4062	1.9172	1.3259	10	4.6	2.1703	1.7854	1.2685
9	8.0	2.2953	1.8452	1.2901	10	4.5	2.0103	1.6629	1.1987
9	4.5	2.2916	1.8498	1.2906	10	4.4	1.8783	1.5643	1.1426
9	4.4	2.2605	1.8398	1.2795	10	4.3	1.7891	1.4991	1.1031
9	4.3	2.1868	1.7877	1.2470	10	4.2	1.7306	1.4553	1.0761
9	4.2	2.0775	1.7014	1.1963	10	4.1	1.6958	1.4282	1.0602
9	4.1	1.9429	1.5951	1.1352	10	4.0	1.6819	1.4184	1.0537
9	4.0	1.7967	1.4838	1.0710	10	3.3	1.6663	1.4298	1.0557
9	3.9	1.6754	1.3937	1.0186	10	3.2	1.6331	1.4304	1.0499
9	3.8	1.5929	1.3335	0.9821	10	3.1	1.6103	1.4053	1.0384
9	3.7	1.5377	1.2926	0.9565	10	3.0	1.7191	1.3563	1.0142
9	3.6	1.5038	1.2671	0.9401	10	2.9	1.6053	1.2841	0.9699
9	3.5	1.4892	1.2581	0.9356	10	2.8	1.4678	1.2062	0.9194
9	3.0	1.5013	1.2555	0.9376	10	2.7	1.3716	1.1404	0.8769
9	2.9	1.4795	1.2632	0.9354	10	2.6	1.3166	1.0963	0.8477
9	2.8	1.4509	1.2590	0.9273	10	2.5	1.2748	1.0757	0.8319
9	2.7	1.4201	1.2256	0.9090	10	2.4	1.2332	1.0826	0.8237
9	2.6	1.3705	1.1671	0.8746	10	2.3	1.2089	1.0798	0.8188
9	2.5	1.2934	1.0952	0.8287	10	2.2	1.2861	1.0482	0.8126
9	2.4	1.2140	1.0282	0.7850	10	2.1	1.2173	1.0105	0.7948
9	2.3	1.1582	0.9783	0.7530	10	2.0	1.1315	0.9640	0.7584
9	2.2	1.1237	0.9502	0.7330	10	1.9	1.0471	0.9209	0.7147
9	2.1	1.0937	0.9442	0.7230	10	1.8	0.9836	0.8784	0.6803
9	2.0	1.0722	0.9418	0.7167	10	1.7	0.9647	0.8431	0.6585
9	1.9	1.0770	0.9157	0.7061	10	1.6	0.9532	0.8128	0.6457
9	1.8	1.0350	0.8743	0.6805	10	1.5	0.8932	0.7832	0.6202
9	1.7	0.9609	0.8219	0.6434	10	1.4	0.8273	0.7351	0.5809
9	1.6	0.8927	0.7758	0.6063	10	1.3	0.7779	0.6873	0.5465
9	1.5	0.8448	0.7422	0.5789	10	1.2	0.7396	0.6478	0.5194
9	1.4	0.8164	0.7127	0.5589	10	1.1	0.6862	0.6066	0.4879
9	1.3	0.7774	0.6758	0.5350	10	1.0	0.6284	0.5584	0.4505
9	1.2	0.7190	0.6299	0.5006	10	0.9	0.5758	0.5122	0.4150
9	1.1	0.6646	0.5852	0.4659	10	0.8	0.5198	0.4635	0.3779
9	1.0	0.6165	0.5445	0.4351	10	0.7	0.4616	0.4127	0.3379
9	0.9	0.5645	0.4990	0.4015	10	0.6	0.4017	0.3606	0.2966
9	0.8	0.5084	0.4510	0.3644	10	0.5	0.3398	0.3058	0.2526

Table 1 (continued)

n	$\hat{\theta}/T$	α=0.1	α=0.05	α=0.01
10	0.4	0.2758	0.2488	0.2067
10	0.3	0.2094	0.1893	0.1579
10	0.2	0.1407	0.1273	0.1064
10	0.1	0.0704	0.0637	0.0532

Table 2 gives values of θ_*/T calculated iteratively, by means of the chi-square approximation in conjunction with the Wilson-Hilferty approximation, for selected values of $\hat{\theta}/T \leq 5$, $n = 40$, $\alpha = 0.1$ and 0.05. Here, as well as for the values of $\hat{\theta}/T$ and θ_*/T covered in Table 1 for a fixed combination of α and n, one can interpolate by converting to $\ell n(\hat{\theta}/T)$ and using linear interpolation to determine a corresponding $\ell n(\theta_*/T)$. An example is given in Section 5.

TABLE 2. Approximate chi-square values of θ_*/T. Lower confidence bound at level $1 - \alpha$, corresponding to $\hat{\theta}/T$, for $n = 40$, $\alpha = 0.1$, 0.05.

$\hat{\theta}/T$	θ_*/T	
	$\alpha = 0.1$	$\alpha = 0.5$
5.0	3.065	2.824
4.0	2.591	2.396
3.0	2.057	1.912
2.0	1.460	1.366
1.0	0.783	0.739
0.5	0.407	0.385
0.2	0.166	0.157
0.1	0.083	0.078

We are at present investigating methods combining simulation techniques and smoothing procedures, so that tabulations can be made over the range $1 < \hat{\theta}/T < n$ for $n > 10$. Results of these investigations will appear in a later paper.

4. Tabulation of the Confidence Bounds

Values of θ_*/T calculated by means of the computer program appear in Table 1 for $n = 2(1)10$, $\alpha = 0.1$, 0.05 and 0.01. These have been checked for accuracy, as described in Section 3, and should be correct to within a unit in the second significant figure. Some values have four significant figures of accuracy.

The range of values of $\hat{\theta}/T$ exhibited in Table 1 reflects the range of values that this estimator is able to take on. For example, if $n = 9$ and the number of failures r is equal to 1, then, from (1), one sees that $\hat{\theta}$ is equal to 8T, plus an increment that ranges from zero to slightly less than T. If $r = 2$, then $\hat{\theta}$ must be less than 4.5T, but no less than 3.5T. If $r \geq 3$, $\hat{\theta}$ may range from zero to 3T. In fact, for $n = 2$, $0.0 < \hat{\theta}/T < 2.0$; for $n = 3(1)6$, $0 < \hat{\theta}/T < n/2$ or $n - 1 < \hat{\theta}/T < n$; and for $n = 7(1)10$, $0 < \hat{\theta}/T < n/3$ or $n/2 - 1 < \hat{\theta}/T < n/2$ or $n - 1 < \hat{\theta}/T < n$.

5. An Example

Consider exponential failure times generated by a sample of size 5 of electronic parts that have been "burnt in". Here, T is equal to 106 days and there are four observed failure times, namely, 1.2, 19.6, 45.1 and 91.3 days, so that $\hat{\theta} = [1.2 + 19.6 + 45.1 + 91.3 + 106]/4 = 65.8$ days and $\hat{\theta}/T = 0.621$.

To obtain a 95 percent lower confidence bound for θ, we first calculate $\ell n(0.6) = -0.5108$, $\ell n(0.7) = -0.3567$ and $\ell n(0.621) = -0.4764$. Then, interpolation in Table 1 yields for $\ell n(\theta_*/T)$, $\ell n(0.2803) + 0.2237(\ell n(0.3228) - \ell n(0.2803))$ $= -1.2719 + 0.2237(-1.1307 + 1.2719) = -1.2403$. Thus, θ_*/T is equal to 0.289, and 30.7 days is a 95 percent lower confidence bound for θ.

ACKNOWLEDGEMENTS

Research performed by Dr. Mann and Mr. Han was supported by the U.S. Office of Naval Research under Contract Nos. N00014-80-C-0684 and N00014-82-K-0023.

REFERENCES

Bain, L.J., Engelhardt, M. and Wright, F.T. (1977). Inferential procedures for the truncated exponential distribution. Communications in Statistics - Theory and Methods A6(2), 103-111.

Barlow, R.E., Madansky, A., Proschan, F. and Scheuer, E.M. (1968). Statistical estimation procedures for the "burn-in" process. Technometrics 10, 51-62.

Bartholomew, D.J. (1957). A problem in life testing. Journal of the American Statistical Association 52, 350-355.

Bartholomew, D.J. (1963). The sampling distribution of an estimate arising in life testing. Technometrics 5, 361-374.

Bartlett, M.S. (1953a). Approximate confidence intervals. Biometrika 40, 12-19.

Bartlett, M.S. (1953b). On the statistical estimation of mean life-times. Philosophical Magazine, Series 7, 44, 249-262.

Deemer, W.L., Jr. and Votaw, D.F., Jr. (1955). Estimation of parameters of truncated or censored exponential distributions. Annals of Mathematical Statistics 26, 498-504.

Halperin, M. (1950). Estimation in truncated sampling processes. RM-370, The Rand Corporation (ASTIA number ATI 210639).

Halperin, M. (1960). Extension of the Wilcoxon-Mann-Whitney test to samples censored at some fixed point. Journal of the American Statistical Association 55, 125-138.

Hoem, J.M. (1969). The sampling distrubution of an estimator arising in connection with the truncated exponential distribution. Annals of Mathematical Statistics 40, 702-703.

Mann, N.R., Schafer, R.E. and Singpurwalla, N.D. (1974). Methods for Statistical Analysis of Reliability and Life Data. New York, John Wiley.

McGinnis, D.F. and Sammons, W.H. (1970). Daily streamflow simulation. Journal of Hydraulics Division, Proceedings of the American Society of Civil Engineers, 1201-1206.

Patnaik, P.B. (1949). The noncentral χ^2 and F distributions and their applications. Biometrika **36**, 202-232.

Severo, N.C. and Zelen, M. (1960). Normal approximation to the chi-square and non-central F probability functions. Biometrika **47**, 411-416.

Spurrier, J.D. and Wei, L.J. (1980). A test of the parameter of the exponential distribution in the type 1 censoring case. Journal of the American Statistical Association **75**, 405-409.

Yang, G. and Sirvanci, M. (1977). Estimation of a time-truncated exponential parameter used in life testing. Journal of the American Statistical Association **72**, 444-447.

SCREEN TESTING AND CONDITIONAL PROBABILITY OF SURVIVAL

Janet Myhre

Claremont McKenna College, Claremont, California

and

Sam Saunders

Washington State University, Pullman, Washington

1. Introduction

There have been only a few parametric models extensively examined for application to reliability; these include the exponential distribution of Epstein-Sobel (1953) and the Weibull distribution (1961). The one most widely utilized for electronic components has been the exponential model, not only because of its simple and intuitive properties but also because of the extent of the estimation and sampling procedures which have been developed from the theory. However, neither of these models is applicable to the study of screen testing.

One of the early discoveries was that mixtures of exponentially distributed random variables have a decreasing failure rate (Proschan, 1963). Thus any two groups of components with constant, but different, failure rates would, if mixed and sampled at random, exhibit a decreasing failure rate. As a consequence, the family of life lengths with decreasing failure rate certainly arises in practice and particular subsets of this family could be of great utility for specific applications, see, e.g., Cozzolino (1968). We examine one such model with shape and scale parameters α and β, respectively, which is based upon a gamma mixture of exponential distributions. This family was introduced by Afanas'ev (1940) and later by Lomax (1954) as a generalization of a Pareto distribution.

A very important property of this gamma-mixed exponential distribution is that the conditional life remaining after a time τ is again distributed as a gamma-mixed exponential. This is shown in Section 3 and its application to screen testing is discussed.

Kulldorff and Vännman (1973) and Vännman (1976) have studied a variant of the gamma-mixed exponential model containing a location parameter. They obtained a best linear unbiased estimate of the scale parameter assuming that the shape parameter α was known and in a region restricted so that both the mean and the variance exist, namely $\alpha > 2$. When this restriction of $\alpha > 2$ cannot be met, an estimate based on a few order statistics, which are optimally spaced, is given, and tables of the weights as functions of the number of spacings are provided. The estimate based on these order statistics for the $\alpha > 2$ case is claimed to be an asymptotically best linear unbiased estimate. In all cases, the shape parameter was assumed known and the sample was either complete or Type II censored. It is contended that BLUE estimates of the shape parameter are not attainable.

Harris and Singpurwalla (1968) examined the method of moments as an estimation procedure for this same model but again with the shape parameter restricted to $\alpha > 2$ and with a complete sample.

Harris and Singpurwalla (1969) also exhibited the maximum likelihood equations but only for complete samples and without resolving the question of existence of solutions. In Kulldorff and Vännman (1973) there is a brief bibliography of results on parametric estimation for this distribution under various assumptions.

As a consequence of the widespread adoption of integrated circuitry, life testing in electronic manufacturing virtually always provides incomplete samples. This is because of censored tests, the expense and the paucity of failures owing to the high reliability of integrated circuit devices. Such service life data cannot be adequately treated by any of the presently known statistical techniques without employing Bayesian arguments, with their utilization of subjective information. This would indicate the need for an objective estimation procedure making use of the only type of data available and without the potential for bias inherent in Bayesian priors.

In this paper the maximum likelihood estimates are obtained for both the shape and scale parameters of the gamma-mixed exponential (Lomax distribution), and sufficient conditions are given for their existence. These estimates are derived for censored data and *a fortiori* for complete samples, even with a paucity of failure observations.

The existence conditions obtained here for the maximum likelihood estimates apply even to the case where the variance and possibly the mean do not exist: $0 < \alpha \leq 2$. Moreover, the estimates of the shape parameter α which have been obtained from actual data indicate that this region $0 < \alpha \leq 2$ is important because all the estimates obtained of α have been less than unity.

2. The Model

We postulate that the underlying process which determines the length of life of a component under consideration is the following: The quality of construction determines a level of resistance to stress which the component can tolerate. The service environment provides shocks of varying magnitude to the component, and failure takes place when, for the first time, the stress from an environmentally induced shock exceeds the strength of the component.

If the time between shocks of any magnitude is exponentially distributed with a mean depending upon that magnitude, then the life length of each component will be exponentially distributed with a failure rate which is determined by the quality of assembly. It follows that each component has a constant failure rate but that the variability in manufacture and inspection techniques forces some components to be extremely good while a few others are bad and most are in-between.

Let X_λ be the life length of a component in such a service environment, with a constant failure rate λ which is unknown. The variability of manufacture determines various percentages of the λ-values and this variability can be described by some distribution, say G.

Let T be the life length of one of the components which is selected at random from the population of manufactured components. We denote the reliability of this component by R and we have

$$R(t) = P[T > t] \quad \text{for } t > 0 \ .$$

Let Λ be the random variable which has distribution G. We can write

$$(1) \qquad R(t) = E\, P[X_\lambda > t \mid \Lambda = \lambda] = \int_0^\infty e^{-\lambda t}\, dG(\lambda) \ .$$

Because of having a form which can fit a wide variety of practical situations when both scale and shape parameters are disposable, it is assumed that G is a gamma distribution, i.e., for some $\alpha > 0$, $\beta > 0$,

$$g(\lambda) = \frac{\lambda^{\alpha-1} e^{-\lambda/\beta}}{\Gamma(\alpha)\beta^\alpha} \quad \text{for } \lambda > 0 \ .$$

That this assumption is robust, even when mixing as few as five equally weighted λ's, has been shown by Sunjata (1974) in an unpublished thesis. It follows from equation (1) that the reliability function is

$$(2) \qquad R(t) = \frac{1}{(1 + t\beta)^\alpha} = e^{-\alpha \ln(1 + t\beta)} \ .$$

The failure rate, hazard rate, can be shown to be

$$(3) \qquad q(t) = \frac{\alpha\beta}{1 + t\beta} \ ,$$

which is a decreasing function of $t > 0$.

Maximum likelihood estimates for α, β and hence $R(t)$ and $q(t)$ are given in Section 4.

3. Residual Life Property of the Model

An important property of this model is that residual life on a component is distributed as a gamma-mixed exponential. Thus a "burn-in" test of a component will yield a residual life which is also in the same family. This property also holds for mixed exponential distributions where the mixing distribution is other than gamma.

The residual life T_h of a component is defined to be the life remaining after time h, given that the component is alive at time h. It can be shown that: A burn-in for h units of time on a component with initial life determined by a gamma-mixed exponential distribution with parameters α and β will yield a residual life T_h and will be distributed as a gamma-mixed exponential with parameters α and $\beta_h = \frac{\beta}{1+\beta h}$.

It follows that this life length model is "used better than new" or "new worse than used" in the sense that T_h is stochastically larger than T for all $h > 0$.

An important consequence of this property is that one can calculate the value of the increased reliability attained by burn-in procedures as compared with the cost of conducting them. It has long been the practice to burn-in electronic components based on intuitive ideas of "infant mortality" in order to provide reasonable assurance of having detected all defectively assembled units. This model, whenever it is applicable, makes possible an economic analysis. A variation of this result has been discussed by Bhaltacharya (1963).

Consider the following data from a screen test of flight control electronic packages:

Failure times:	1, 8, 10
Alive times:	59, 72, 76, 113, 117, 124, 145, 149,
	153, 182, 320 .

Each package has recorded, in minutes, either a failure time or an alive time. An alive time is the time the test was terminated with the package still functioning. In this example the reason for the censoring was that the test equipment was needed elsewhere, that funds for testing were depleted, etc. We define this censoring to be *random*, that is, when a package is operating the time of censoring depends on some censoring distribution which is independent of the failure distribution. Type I censoring at a preassigned time is a special case of random censoring. Random censoring differs, however, from Type II censoring, where censor occurs at some preassigned number of failures. For this example the censoring is random in that time on test equals minimum {failure time, random censor time}.

The example data do not exhibit a constant failure rate. Even if we assume a fourth failure rate at 59 minutes and Type II censoring at the fourth failure, we reject the hypothesis of a constant or increasing failure rate in favor of a decreasing failure rate (using F-criteria suggested by Gnedenko, et al. (1969)). If we assume that the data are from a gamma-mixed exponential, we find (using equations in Section 4) that $\hat{\alpha} = 0.0453$ and $\hat{\beta} = 1.03$.

The question is: how long should packages of this type be subjected to a screen test? Let $\hat{\beta}_h$ denote the estimated scale parameter after a burn-in for h units of time. For this package 96 hours of burn-in time has been used. Figure 1 shows $\hat{\beta}_h$ as a function of h, with $\hat{\beta}$ as above. If it could be assumed that burn-in tests were equivalent to actual use tests, then one could estimate reliability at time t as a function of burn-in time h. For example

$$\hat{R}(t) = [1 + t\hat{\beta}_h]^{-\hat{\alpha}} \quad .$$

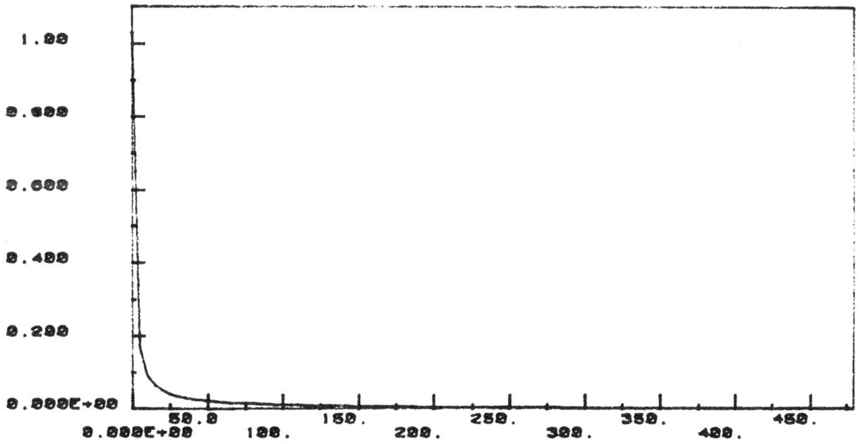

FIGURE 1. Graph of $\hat{\beta}_h$ as a function of burn-in time h in minutes.

Figure 2 shows the change in estimated reliability at 20 minutes as a function of burn-in time h, in minutes, for $\hat{\alpha}$ and $\hat{\beta}$ as above.

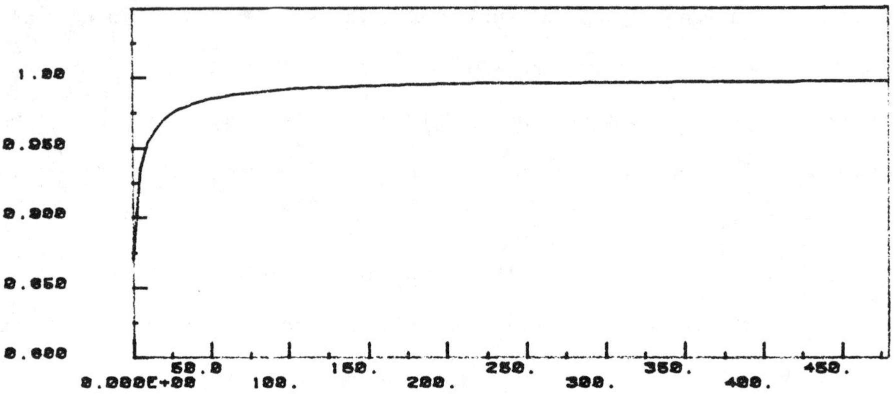

FIGURE 2. Graph of estimated burn-in reliability at 20 minutes as a function of previous burn-in time h in minutes.

Since burn-in reliability is usually proportional to in-use reliability, one must question accepted burn-in times of 96 hours for equipment which exhibit burn-in data of the type given in the example.

4. Estimation of Parameters Using Incomplete Samples

When components having a decreasing failure rate are tested, the samples are virtually always incomplete in the sense that testing is stopped before all components have failed. A datum on a component that "failure has not yet occurred after a specified life" is called an alive time. Samples containing such observations are censored. In our experience with electronic components, Type I and Type II censoring occur less frequently than random censoring. However, until recently Type I and Type II censoring were the cases addressed in the literature. The results which follow apply to Type I, Type II, or random censoring, as defined above. Complete samples are a special case.

It is assumed in this section that we are given a sample $\underline{t} = (t_1,\ldots,t_k,\ldots, t_n)$, where $t_1,\ldots,t_k$ are ordered observations of times of failures while $t_{k+1},\ldots,t_n$ are ordered observed alive times where censoring is one of the types described above. It is assumed that $k \geq 1$.

Some results will now be given on maximum likelihood estimation of the unknown shape and scale parameters for the gamma-mixed exponential in the case of censored samples.

Remark 1: When the scale parameter β is known, there exists a m.l.e. of α given by

$$\hat{\alpha} = \frac{k}{\sum_{1}^{n} \ell n(1+\beta t_i)} .$$

This result is not new. If β is known then the values $y_i = \ell n(1+\beta t_i)$ are observations from an exponential distribution with unknown failure rate α.

THEOREM 2: For given $\alpha, \underline{t}$ there exist a unique m.l.e. of β, denoted by $\hat{\beta}$. It is given implicitly as the positive root of the equation

$$(4) \qquad 1 - \frac{1}{k} \sum_{1}^{k} \frac{\beta t_j}{1+\beta t_j} - \frac{\alpha}{k} \sum_{1}^{n} \frac{\beta t_i}{1+\beta t_i} = 0 .$$

PROOF: We have the vector $\underline{t} = (t_1, \ldots, t_k, \ldots, t_n)$ corresponding to the observed events $[T_i = t_i]$ for $i = 1, \ldots, k$ and $[T_i > t_i]$ for $i = k+1, \ldots, n$, where $k \geq 1$. By definition the log-likelihood is given by

$$e^L = \prod_{j=1}^{k} q(t_j) \prod_{i=1}^{n} R(t_i) .$$

Substituting, taking logarithms, and simplifying we have

$$L(\beta|\alpha,\underline{t}) = k\ell n(\alpha\beta) - \sum_{j=1}^{k} \ell n(1+\beta t_j) - \alpha \sum_{i=1}^{n} \ell n(1+\beta t_i) .$$

Dividing by k we write

$$L(\beta|\alpha,\underline{t}) = \ell n\alpha + \ell n\beta - \frac{1}{k} \sum_{j=1}^{k} \ell n(1+\beta t_j) - \frac{\alpha}{k} \sum_{i=1}^{n} \ell n(1+\beta t_i) \cdot$$

$$L'(\beta|\alpha,\underline{t}) = \frac{1}{\beta} \left[1 - \frac{1}{k} \sum_{j=1}^{k} \frac{\beta t_j}{1+\beta t_j} - \frac{\alpha}{k} \sum_{i=1}^{n} \frac{\beta t_i}{1+\beta t_i} \right] .$$

For $k \geq 1$, the m.l.e. for β is given as the positive root of

(6) $$A(\beta) = 1 - \frac{1}{k}\sum_{j=1}^{k}\frac{\beta t_j}{1+\beta t_j} - \frac{\alpha}{k}\sum_{i=1}^{n}\frac{\beta t_i}{1+\beta t_i} = 0 .$$

The root of equation (6) exists and is unique since $A(0) = 1$, $A(\infty) < 0$ and A is strictly decreasing over $[0,\infty]$.

When α and β are both unknown there are sets of n positive numbers (say $\underline{t} = (t_1,\ldots,t_k,\ldots,t_n)$ with $k \leq n$ designated as alive times) that cannot be used to estimate both unknown parameters. It is shown that both m.l.e.'s exist whenever the sample satisfies the condition

(7) $$2\left(\sum_{j=1}^{k}\frac{t_j}{k}\right)\left(\sum_{i=1}^{n}\frac{t_i}{n}\right) < \sum_{i=1}^{n}\frac{t_i^2}{n} .$$

If the sample fails to satisfy this condition then the model may not be appropriate and a constant failure rate model or a convex failure rate model may be indicated rather than a decreasing failure rate model.

One can check that a complete sample of failure times, i.e., with $k = n$ will satisfy (7) if the sample standard deviation exceeds the mean. For decreasing failure rate distributions the standard deviation does exceed the mean, when they both exist.

THEOREM 3: For a given sample $\underline{t}$, with $k \geq 1$, both α,β unknown, the m.l.e. of $\beta,\hat{\beta}$, exists as the smallest positive root of

(8) $$1 - \frac{1}{k}\sum_{j=1}^{k}\frac{\beta t_j}{1+\beta t_j} - \frac{\displaystyle\sum_{i=1}^{n}\frac{\beta t_i}{1+\beta t_i}}{\displaystyle\sum_{i=1}^{n}\ln(1+\beta t_i)} = 0 ,$$

if the sample satisfies

$$2\left(\sum_{i=1}^{n} \frac{t_i}{n}\right)\left(\sum_{j=1}^{k} \frac{t_j}{k}\right) < \sum_{i=1}^{n} \frac{t_i^2}{n} .$$

Given $\hat{\beta}$, the m.l.e. of α is given by

$$\hat{\alpha} = \frac{k}{\sum_{i=1}^{n} \ell n(1+\hat{\beta}t_i)} .$$

PROOF: Consider the log-likelihood function defined in (5). Since α is unknown, we write (5) as $L(\alpha,\beta|\underline{t})$. All stationary points, which are determined by $\underline{t}$, can be found by the simultaneous solution of $\partial L(\alpha,\beta|\underline{t})/\partial\alpha = 0$ and $\partial L(\alpha,\beta|\underline{t})/\partial\beta = 0$. This yields two equations in α and β:

(9) $$\frac{1}{\alpha} = \frac{1}{k}\sum_{i=1}^{n} \ell n(1+\beta t_i); \quad 1 - \frac{1}{k}\sum_{j=1}^{k} \frac{\beta t_j}{1+\beta t_j} = \frac{\alpha}{k}\sum_{i=1}^{n} \frac{\beta t_i}{1+\beta t_i} .$$

Combining these into a single equation in β, we seek $\hat{\beta}$ as the root of equation (8), i.e., the root of

$$B(\beta) = 1 - \frac{1}{k}\sum_{j=1}^{k} \frac{\beta t_j}{1+\beta t_j} - \frac{\sum_{i=1}^{n} \frac{\beta t_i}{1+\beta t_i}}{\sum_{i=1}^{n} \ell n(1+\beta t_i)} = 0 .$$

Note that $\lim_{\beta\to 0} B(\beta) = 0$. We want to find a sufficient condition for the equation $B(\beta) = 0$ to have a positive root. It is clear that $B(\beta) \underset{\beta\to\infty}{\to} 0$ as a negative quantity. We will show that if the sample satisfies condition (7) then $B(\beta)$ is positive in a neighborhood of zero. Under these conditions there exists a β such that $B(\beta) = 0$.

To show that if condition (7) holds then $B(\beta)$ is positive in a neighborhood of zero consider the function

$$\frac{1}{\beta} B(\beta) = -\left[\frac{\sum_{i=1}^{n} \frac{t_i}{1+\beta t_i} - \frac{1}{k} \sum_{j=1}^{k} \frac{1}{1+\beta t_j} \sum_{i=1}^{n} \frac{\ell n(1+\beta t_i)}{\beta}}{\sum_{i=1}^{n} \ell n(1+\beta t_i)}\right] .$$

Repeated application of l'Hôpital's rule shows that

$$\lim_{\beta \to 0} \frac{1}{\beta} (B(\beta)) = \frac{\sum_{i=1}^{n} t_i^2/2 - \frac{1}{k} \sum_{j=1}^{k} t_j \sum_{i=1}^{n} t_i}{\sum_{i=1}^{n} t_i} ,$$

which is positive if $\sum_{i=1}^{n} \frac{t_i^2}{n} > 2 \sum_{j=1}^{k} \frac{t_j}{k} \sum_{i=1}^{n} \frac{t_i}{n}$.

The solution for $\hat{\alpha}$ follows immediately from the first equation in (9).

Computationally equations (4) and (8) are not difficult to solve. In fact, their solutions are obtainable using a simple programmable calculator.

5. Conclusion

If screen tests are effective we should be observing a decreasing failure rate as a function of time on test. In practice it is often assumed that as the result of screen tests the surviving components are exponentially lived. Of course, this is not always the case. The important question is how long should a component be burned-in in order to make its residual life distribution acceptable? The decreasing failure rate gamma-mixed exponential model, when applicable, allows estimation of the improvement in reliability as a function of screen test time.

This study suggests that if a component has a life distribution with decreasing failure rate it is the alive times within the data which contribute principally to the estimation of the parameters (and thereby to the determination of reliability) since only one failure observation is required even to estimate two parameters, presuming the data of alive times are ample.

The usual justification for using maximum likelihood estimates is owing to their asymptotically optimal properties and to their asymptotic normality. The problem of obtaining exact sampling distributions of the maximum likelihood estimators of the parameters for the model studied seems to be difficult because the estimates are only implicitly defined. Myhre and Saunders (1981) have shown that when they exist, the m.l.e.'s for α and β based on Type I or on random sampling are consistent and are asymptotically normally distributed. In addition, Lucke and Myhre (1980) have shown that the distribution function estimated using the joint m.l.e.'s of the parameters is closer to the true distribution function for regions of interest in reliability theory than is the estimated distribution function using a known shape parameter and the BLUE estimate of Vännman (1976) for the scale parameter.

ACKNOWLEDGEMENTS

This research was supported in part by the Office of Naval Research Grants N00014-78-C-0213 and N00014-79-C-0755.

REFERENCES

Afanas'ev, N.N. (1940). Statistical theory of the fatique strength of metals. Zhurnal Technicheska Fiziki **10**, 1553-1568.

Bhaltacharya, N. (1963). A property of the Pareto distribution. Sankhya B **25**, 195-196.

Cozzolino, J.M. (1968). Probabilistic models of decreasing failure rate processes. Naval Research Logistic Quarterly **15**, 361-374.

Epstein, B. and Sobel, M. (1953). Life testing. Journal of the American Statistical Association **48**, 486-502.

Gnedenko, B, et al (1969). Mathematical methods of reliability theory. Academic Press, New York.

Harris, C.M. and Singpurwalla, N.D. (1968). Life distributions derived from stochastic hazard functions. IEEE Transactions in Reliability, **R-17**, 70-79.

Harris, C.M. and Singpurwalla, N.D. (1969). On estimation in Weibull distributions with random scale parameters. Naval Research Logistic Quarterly **16**, 405-410.

Kulldorff, G. and Vännman, K. (1973). Estimation of the location and scale parameters of a Pareto distribution. Journal of the American Statistical Association **68**, 218-227.

Langberg, N.A., Lión, R.V., Proschan, F., Lynch, J. (1978). The extreme points of the set of decreasing failure rate distributions. FSU Technical Report.

Lomax, K.S. (1954). Business failures: another example of the analysis of failure data. Journal of the American Statistical Association **49**, 847-853.

Lucke, J. and Myhre, J. (1980). A comparison between MLE and BLUE for a decreasing failure rate distribution when using censored data. Unpublished report, Institute of Decision Science, Claremont McKenna College.

Myhre, J. and Saunders, S. (1981). On problems of estimation of two parameter decreasing failure rate distributions applied to reliability. Unpublished report, Institute of Decision Science, Claremont McKenna College.

Proschan, F. (1963). Theoretical explanation of observed decreasing failure rate. Technometrics **5**, 375-383.

Sunjata, M.H. (1974). Sensitivity analysis of a reliability estimation procedure for a component whose failure density is a mixture of exponential failure densities. Naval Postgraduate School, Monterey. Unpublished thesis.

Vännman, K. (1976). Estimators based on order statistics from a Pareto distribution. Journal of the American Statistical Association **71**, 704-708.

Weibull, W. (1961). Fatique testing and analysis of results. Pergamon Press, New York.

IMPERFECT MAINTENANCE

Mark Brown

City University of New York

and

Frank Proschan

Florida State University

1. Introduction

An impressive array of mathematical and statistical papers and books have appeared in which a variety of maintenance policies are studied to determine their performance and to achieve optimization. In most of the models treated, it is assumed that the relevant information to be used is available and correct, and that maintenance actions are carried out as specified in the maintenance policy being used or to be used.

Unfortunately, the most important factor in a great many actual maintenance operations is omitted, thus vitiating the solution theoretically determined.

The most important factor, inadvertently overlooked or deliberately ignored for the sake of mathematical tractability, is the fallibility of the maintenance performer. In actual practice (as contrasted with the model formulation), the maintenance performer may:

(1) Repair the wrong part.

(2) Only partially repair the faulty part.

(3) Repair (partially or completely) the faulty part, but damage adjacent parts.

(4) Incorrectly assess the condition of the unit inspected.

(5) Perform the maintenance action not when called for, but at his convenience.

Clearly, the list of imperfect and even destructive repair actions occurring in actual maintenance may be extended much further. Apparently, we need mathematical models for maintenance which take explicit account of imperfect and/or destructive repair and faulty inspection.

In this paper, we formulate a variety of more realistic models which incorporate explicitly imperfect maintenance actions, postulating a probabilistic basis for their occurrence. We do not attempt to carry out solutions to the problems arising within these models, unselfishly leaving this challenging and enjoyable (?) task to the eager doctoral candidate looking for a dissertation topic and the young professor searching for problem areas yielding publications so vital for tenure and promotion. We do, however, summarize the main results of one study we carried out in detail of the type described above.

Throughout we assume a life distribution of F(t) for the unit being maintained.

2. Planned Replacement Based on Time Elapsed

First we succinctly describe three basic maintenance models which have been widely used. (See Barlow and Proschan, 1965, Chapters 3 and 4). In these classical models, no provision has been made for imperfect maintenance. We then describe a variety of imperfect repair or inspection actions on a probabilistic basis with their resultant adverse effects and costs. Some or all of these difficulties may be incorporated into the three models as originally formulated to achieve more realistic descriptions of maintenance situations as they actually occur in practice.

2.1 Age Replacement

A unit is replaced at failure or at age T (constant), whichever comes first. Operation and replacement continue indefinitely in this fashion. Given the cost c_1 of failure during operation and the cost $c_2(< c_1)$ of planned replacement, describe the operating characteristics of this policy and determine the optimal value of T, i.e., the value minimizing the long-run cost per unit of time. Compute the resulting minimal cost per unit of time in the long run.

2.2 Block Replacement

A unit is replaced at fixed times T, 2T,..., . In addition, it is replaced at failure. The cost of failure replacement is c_1, while the cost of a planned replacement is $c_2(< c_1)$. Describe the operating characteristics of the policy and find the value of T minimizing the long run cost per unit of time. Compute the resulting minimum long run cost per unit of time.

In both the age replacement and block replacement policies, replacement of a failed unit is by a new unit. Thus the instant of replacement represents a regeneration point in the stochastic process.

2.3 Maintenance with Minimal Repair

A unit is replaced by a new unit at fixed time T. If failure occurs at time $t < T$, minimal repair occurs, returning the unit to the functioning state, but the condition of the unit is that of a unit of age t Any number of failures resulting in minimal repair may occur during [0,T]. The cost of minimal repair is c_2; the cost of planned replacement by a new unit at time T is c_1, and $c_1 > c_2$. As before, derive the operating characteristics of the policy, find the optimizing value of T, and compute the resulting minimal cost per unit of time in the long run.

We now formulate imperfect maintenance features on a probabilistic basis. These may be incorporated into any of the three basic models above to achieve a more realistic description of maintenance as it occurs in practice.

2.4 Imperfect Maintenance Features

(a) Planned replacement may occur not at the time or age planned but may deviate from this time by an amount D which is governed by a probability distribution F_D.

(b) The replacement is not always installed properly. With probability p it is installed properly, and with probability $q = 1 - p$ it is faultily installed; the resulting failure rate function $r_I(t) = ar(t)$, where $r(t)$ is the failure rate function corresponding to proper installation and $a > 1$.

(c) With probability d, $0 < d < 1$, failure of a unit is detected immediately; with probability density $g(t)$, failure of a unit is detected t units of time following its occurrence, $t > 0$. A cost of $c_3 t$ is incurred if failure remains undetected for t units of time. This cost is in addition to the cost c_1 of failure replacement.

(d) With probability q', minimal repair at time t_0 may be imperfectly performed, leading to a functioning unit of effective age $t_0 + m$, rather than the actual age t_0, $m > 0$. Each time a minimal repair is imperfectly performed, the repaired unit's effective age increases by m units of time.

(e) In addition to the random features corresponding to imperfect maintenance, information as to costs may be uncertain. Thus c_1, c_2, c_3 may be random, governed by distributions F_1, F_2, F_3, respectively.

3. Maintenance Based on Inspection

The models of Section 2 called for maintenance actions based on unit age or chronological time. Another class of maintenance policies calls for maintenance actions based on the physical condition of the unit. Of course, these policies require inspection of the unit according to some plan.

3.1 Periodic Inspection

The simplest plan of inspection is to inspect the unit periodically at interval I, say. For simplicity, we assume that the purpose of the inspection is to determine whether the unit is functioning or failed; a failed unit would remain undetected unless specifically inspected. (An example is a battery serving as a spare; another example is a fire detection device. In the biological realm, the occurrence of cancer in the early stages is an example).

Suppose the cost of each inspection is c_I and the cost of undetected failure is c_U per unit of time between failure and its detection. We wish to determine the optimal value of I (periodic inspection interval) to minimize total expected costs.

3.2 Inspection Interval Dependent on Age

For an item with increasing failure rate, it would be reasonable to schedule inspections more and more frequently as the age of the item increased. Thus, inspections might be scheduled at ages $x_1 < x_2, \ldots,$ where $(x_2 - x_1) > (x_3 - x_2) > \cdots$. From a knowledge of c_I, c_U, and the failure rate function $r(t)$ of the unit, we wish to find the values of $x_1, x_2, \ldots,$ which minimize total expected costs.

3.3 Identifying Failed Unit(s)

A system fails. We inspect the components in sequence to identify the failed units and then replace them in order to resume system functioning. The reliability of component i is p_i and the cost of inspecting component i is c_i, $i = 1, \ldots, n$. Our aim is <u>not</u> to identify all failures (and, of course, to replace them), but only a set sufficient to permit the resumption of system functioning. For example, consider the following system.

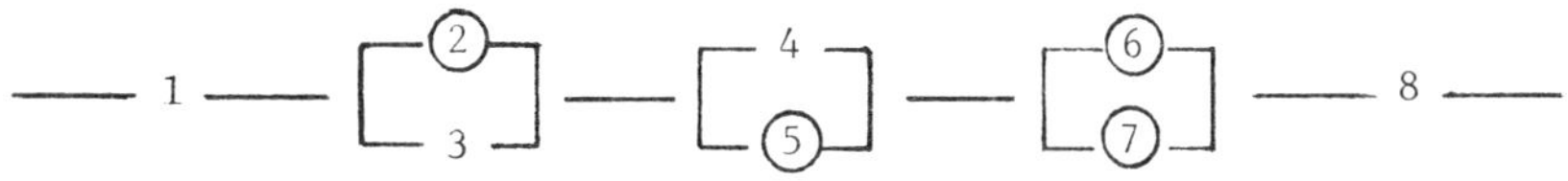

Diagram 1. System diagram showing failed components. Failure is indicated by a circle around component number.

It suffices to identify components 6 and 7 as failed and to replace them to initiate resumption of system functioning.

What is the optimal sequence of component inspection? That is, what sequence of component inspection will result in resumption of system functioning at minimal expected cost?

3.4 Imperfect Inspection Features

(a) Inspection scheduled for time t_0 say, may actually occur at time $t_0 + D$, D random.

(b) Inspection may be imperfect so that with probability q_1 a functioning unit is labeled failed and with probability q_2 a failed unit is labeled functioning. In Model 3.3, these probabilities may differ for different components of the system, so that for component i, these probabilities become q_{1i} and q_{2i}, $i = 1,\dots,n$.

(c) Knowledge of costs may be imperfect, so that the various cost parameters may be considered as random variables.

4. Summary of Results for an Imperfect Repair Model

Brown and Proschan (1980) formulate and analyze an imperfect repair model. They obtain the operating characteristics of the stochastic process generated and monotonicity properties for various parameters and random variables. They do not impose any cost structure nor attempt any optimization. In this section we describe the model and summarize the results obtained. Here we attempt to interpret the results; for mathematical proofs we refer the interested reader to the original paper.

4.1 Model

An item is repaired at failure. With probability p, it is returned to the "good as new" state (perfect repair); with probability $q = 1 - p$, it is returned to the functioning state, but is only as good as an item of age equal to its age at failure (imperfect repair). Repair takes negligible time. The process of alternating failure and repair continues indefinitely over time; we call it a failure process. Finally, suppose the item has underlying life distribution F with failure rate function r(t).

4.2 Stochastic Results Obtained

For the case $p = 0$ (all repairs are imperfect), the failure process is a nonhomogeneous Poisson process with intensity function r(t).

For the more realistic case of $0 < p < 1$, we note that the time points of perfect repair are regeneration points. The interval between successive regeneration points is the waiting time for a perfect repair starting with a new component.

Let F_p denote the waiting time distribution for a perfect repair starting with a new component. Let r_p denote its failure rate function. Then we show:

LEMMA 1: (i) $r_p(t) = pr(t)$. (ii) $\overline{F}_p(t) = \overline{F}^{\,p}(t)$.

Aside from its immediate value in understanding the imperfect repair model, the results of Lemma 1 are interesting in that they represent a physically motivated example of the well-known and widely used model of proportional hazards. The assumption of proportional hazards is often made for mathematical convenience, especially in competing risk theory. Here we see one realistic instance of its occurrence in maintenance theory.

Since the original failure rate function is simply multiplied by p, it follows that many of the important classes of distributions characterized by aging properties are preserved in the following sense: If F has a given aging property, then F_p also has this property for $0 < p < 1$. Formally stated, we have:

THEOREM 1: Let F be in any of the classes: IFR, DFR, IFRA, DFRA, NBU, NWU, DMRL, or IMRL. Then F_p is in the same class.

Theorem 1 cannot be extended to the NBUE and NWUE cases.

Monotonicity Properties

Let $\mu(p)$ denote the mean of F_p. Clearly $\mu(p)$ is a decreasing function. However, the next theorem shows that a reversal in the direction of monotonicity may be achieved by weighting $\mu(p)$ appropriately.

THEOREM 2: (i) Let F_p be NBUE for all p in (0,1]. Then $p\mu(p)$ is increasing for $p \in (0,1]$. In particular, this monotonicity holds for F NBU or DMRL.

(ii) As an immediate consequence of (i), when F_p is NBUE for all p in (0,1], we have the bound:

$$\mu(p) \leq \frac{1}{p} \mu(1) \quad .$$

(iii) Dual results hold for the NWUE, NWU, and IMRL classes.

The inequality $p_1\mu(p_1) \leq p_2\mu(p_2)$ for $p_1 < p_2$ when F_p is NBUE for all $p \in (0,1]$ can be interpreted as: F_{p_1} is smaller in expectation than a geometric sum (with parameter p_1/p_2) of i.i.d. random variables having distribution F_{p_2}. When F is NBU, "smaller in expectation" can be strengthened to "stochastically smaller". A dual result holds for F NWU.

Let $N_p(t)$ denote the number of failures in [0,1) for the failure process in which perfect repair has probability p. We can prove that for F NBU, $N_p(t)$ is stochastically decreasing in p. Intuitively, this is reasonable since greater p leads to a quicker return to the "good as new" state.

A stronger conclusion can be obtained under the stronger assumption of IFR:

THEOREM 3: Let F be IFR. Let Z_p denote the waiting time until the next failure, starting in steady state. Let h_p denote the failure rate function of Z_p. Then (i) for each $t \geq 0$, $h_p(t)$ is decreasing in p; (ii) as a consequence, Z_p is stochastically increasing in p. (iii) Dual results hold for F DFR.

A weaker conclusion is obtained under the weak assumption of F DMRL (IMRL):

COROLLARY 1: Let F be DMRL(IMRL). Then EZ_p is increasing (decreasing) in p.

In all the results above, conclusions were obtained for dual families of distributions corresponding to deterioration with age and improvement with age. The following result applies for DFR distributions, but the dual result is known not to necessarily hold for IFR distributions.

THEOREM 4: Let F be DFR. Let $Z_p(t)$ denote the waiting time at t for the next failure to occur; let $Z_p^*(t)$ denote the waiting time at t for the next perfect repair. Let $m_p(t)$ denote the failure intensity at t and let $m_p^*(t)$ denote the renewal density at t for the renewal process with interarrival time distribution F_p. Finally, let $A_p(t)$ denote the effective age at time t, i.e., the time elapsed since the last perfect repair. Then:

(i) $A_p(t)$, $Z_p(t)$ and $Z_p^*(t)$ are stochastically increasing in t for fixed p; $m_p(t)$ and $m_p^*(t)$ are decreasing in t for fixed p.

(ii) $A_p(t)$ and $Z_p(t)$ are stochastically decreasing in p for fixed t, $m_p(t)$ is increasing in p for fixed t.

ACKNOWLEDGEMENTS

Research performed under the support of the Air Force Office of Scientific Research, AFSC, USAF, under Grant AFOSR F49620-79-C-0157 and Grant AFOSR-81-0038.

REFERENCES

Barlow, R.E. and Proschan, F. (1965). Mathematical Theory of Reliability. Wiley, New York.

Brown, M. and Proschan, F. (1980). Imperfect repair. Florida State University, Department of Statistics Report AFOSR-78-108. Accepted, Journal of Applied Probability.

A LIMIT THEOREM FOR TESTING WITH RANDOMLY CENSORED DATA

Hira L. Koul and V. Susarla

Michigan State University, East Lansing, Michigan

1. Introduction

Let $X_1,\ldots,X_n$ be independent identically distributed random variables (r.v.'s) and $Y_1,\ldots,Y_n$ be independent r.v.'s., independent of $X_1,\ldots,X_n$. Let

$$F(x) = P(X_i > x), \quad x \geq 0$$

$$G_i(y) = P(Y_i > y), \quad y \geq 0, \quad 1 \leq i \leq n \ .$$

Let

$$\delta_i = [X_i < Y_i], \quad Z_i = \min(X_i, Y_i), \quad 1 \leq i \leq n \ .$$

Here, [A] denotes indicator of event A. The X_i's are true survival times, the Y_i's are censoring times and one observes $\{(\delta_i, Z_i),\ 1 \leq i \leq n\}$. This is the so-called random censoring model where often one is interested in making inferences about F or about some function of F based on $\{(\delta_i, Z_i),\ 1 \leq i \leq n\}$. In order to describe the specific problems to be considered here we need the following definitions. In all of these definitions $F(0) = 1$.

DEFINITION 1:

A life distribution $1-F$ is said to be New Better Than Used (NBU) if and only if

(1) $$F(x+y) \leq F(x)\, F(y), \quad x,y \geq 0 \quad .$$

DEFINITION 2:

A life distribution $1-F$ with $0<\mu<\infty$ is said to be Decreasing Mean Residual Life (DMRL) if and only if

(2) $$J(s)\, F(t) \geq J(t)\, F(s), \quad 0 \leq s \leq t < \infty \ ,$$

where

$$J(s) = \int_s^\infty F(t)dt \quad .$$

DEFINITION 3:

A life distribution $1-F$ with mean $0<\mu<\infty$ is said to be New Better Than Used in Expectation (NBUE) if, and only if

(3) $$J(t) \leq \mu\, F(t), \quad t \geq 0 \quad .$$

It is clear that DMRL $\subset$ NBUE. An NBU F with $0 < \mu < \infty$ is also NBUE. An equality obtains in (1) if and only if F is an exponential whereas equality is obtained in (2) and (3) only by an exponential distribution among all continuous F's with $0 < \mu < \infty$.

The probabilistic aspects of the above classes of life distributions have been extensively studied by Bryson and Siddiqui (1969), Bryson (1974), Marshall and Proschan (1972), Barlow and Proschan (1975), among others.

It is of interest, as discussed in Hollander and Proschan (H-P) (1972,1975), Koul (1977,1978a,b) and Koul and Susarla (1980), to test

$$H_0: F(x) = e^{-\lambda x}, \ x \geq 0, \ \lambda > 0 \text{ unknown}$$

against the alternatives

$$H_1: F \text{ is NBU, not an exponential}$$

or

$$H_2: F \text{ is DMRL, not an exponential}$$

or

$$H_3: F \text{ is NBUE, not an exponential} \quad .$$

The papers of Hollander and Proschan and Koul discuss some tests of H_0 vs H_1, H_2 and H_3 when the observations are not censored. The paper of Koul and Susarla discusses a test of H_0 vs H_3 and that of Chen, Hollander and Langberg (1980) discusses a test of H_0 vs H_2 for randomly censored data.

In this paper we present two tests of H_0 vs H_1 and a new test of H_0 vs H_2 for randomly censored data. Besides these the paper contains a limit theorem which is useful in deriving the asymptotic distribution, under H_0 and under alternatives, of the test statistics for the above problems. The theorem partly unifies the proofs of the asymptotic normality of these statistics under random censoring and it also has applications to other problems, such as the estimation of moments of F.

Section 2 contains the main theorem, the tests of H_0 vs H_1 and H_2 based on $\{(\delta_i, Z_i), \ 1 \leq i \leq n\}$, and theorems stating their asymptotic normality along with some proofs. Section 3 has a discussion about the asymptotic Pitman efficiency of some of the tests. Section 4 contains the proof of the main theorem.

NOTATION. The symbols $\sum$ and Π stand, respectively, for the summation and product over the indices $1 \leq i \leq n$. For any function g and set A, $\int_A g$ denotes $\int_A g(x)dx$. All limits are taken as $n \to \infty$. $\bar{G} = n^{-1} \sum G_i$. By $o(1)$ ($o_p(1)$) is meant a sequence of numbers (r.v.'s) that converges to 0 (in probability). z_t denotes the tth percentile of N(0,1) distribution. For any function H, H^{-1} will stand for 1/H. The symbol := stands for "by definition".

2. The Main Theorem and Test Statistics

Let

$$\hat{F}(t) = \frac{1\ N(t)}{1+n} \cdot \Pi \left\{ \frac{1+N(Z_i)}{2+N(Z_i)} \right\}^{[\delta_i = 0,\ Z_i \leq t]}, \quad t \geq 0 \tag{4}$$

denote a modified product limit estimator of F, where

$$N(t) = \sum [Z_i > t], \quad t \geq 0 .$$

Let $\{h_n\}$ be a sequence of non-random functions on $(0,\infty)$ and $\{t_n\}$ be a sequence of positive real numbers, $t_n \uparrow \infty$.

THE MAIN THEOREM:

Let $\{F_n\}$ be a sequence of survival functions and $G_1,\ldots,G_n$ the censoring survival functions. Assume that the following conditions hold:

(C1) $$n^{-1/2}(\ln n)^2 \int_0^{t_n} |h_n(x)| \{\bar{G}(x)\}^{-1} \left(-\int_0^x F_n(H)^{-4}\, d\bar{G}\right) dx = o(1) ,$$

(C2) $$\limsup \sigma_n^2 < \infty ,$$

where $H = F_n \bar{G}$ and

$$(5) \qquad \sigma_n^2 = -\int_0^{t_n} F_n^{-2}(x) \{\bar{G}(x)\}^{-1} \left(\int_x^{t_n} F_n h_n \right)^2 dF_n(x) \quad .$$

Then

$$(6) \qquad n^{1/2} \sigma_n^{-1} \int_0^{t_n} (\hat{F} - F_n) h_n \to_d N(0,1) \quad .$$

Typically $t_n = c(\ell n\, n)^a$, $c > 0$, $0 \leq a < 1$ satisfies (C1) and (C2) for a large class of $\{F_n\}$ and $\{G_i\}$. The proof of this theorem is sketched in Section 4. In this section we now present some important applications of this theorem to the testing problems mentioned in Section 1. First consider the problem of testing:

(a) H_0 versus H_1. Two measures of departure of H_1 from H_0, for a given F, are

$$\Delta_1(F) := \int_0^\infty \int_0^\infty D(s,t)\, ds dt = \int_0^\infty s\, F(s) - \left(\int_0^\infty F\right)^2$$

and

$$\Delta_2(F) := \int_0^\infty \int_0^\infty D(s,t)\, dF(s)\, dF(t) = \int_0^\infty \int_0^\infty F(s+t)\, dF(s)\, dF(t) - 1/4$$

where

$$D(s,t) := F(s+t) - F(s)\, F(t), \quad s,t \geq 0 \quad .$$

The measure Δ_2 was considered by H-P (1972) in the case of no censoring. For some other measures see Koul (1978a,b). Observe that $\Delta_j(F) = 0$, $j = 1,2$ if F is in H_0 and $\Delta_j(F) < 0$, $j = 1,2$ if F is continuous in H_1. The smaller $\Delta_j(F)$ is for a given F, the more there is evidence in favor of F in H_1, $j = 1,2$. Therefore, it is natural to base tests of H_0 vs H_1 on $\Delta_j(\hat{F})$, $j = 1,2$, where $\hat{F}$

is given by (4). Because of the bad tail behavior of $\hat{F}$, we instead consider $\Delta_j(\hat{F},M_n)$, where

$$\Delta_1(F,M_n) = \int_0^{M_n}\int_0^{M_n} D(s,t)\, ds dt \ ,$$

and

$$\Delta_2(F,M_n) = \int_0^{M_n}\int_0^{M_n} D(s,t)\, dF(s)\, dF(t) \quad ,$$

and where $M_n \uparrow \infty$.

The test j rejects for small values of $\Delta_j(\hat{F},M_n)$, $j = 1,2$. The following theorem gives the asymptotic distribution of $\Delta_j(\hat{F},M_n)$, $j = 1,2$ for a sequence $\{F_n\}$ of survival distributions in $H_0 \cup H_1$ and for non-identically distributed censoring r.v.'s. Let $\mu_n = \int_0^\infty F_n$, $\gamma_n = \int_0^\infty sF_n(s)\, ds$. Note that now $X_1, X_2, \ldots, X_n$ are i.i.d. F_n, $n \geq 1$.

THEOREM 2:

(a) Let

$$h_{n1}(x) = (x - 2\mu_n)\,[0 < x < M_n] + (2M_n - x)\,[M_n \leq x < 2M_n] \ .$$

Assume that $\{F_n\}$ in $H_0 \cup H_1$, and $G_1, \ldots, G_n$ satisfy (C1) and (C2) with $h_n = h_{n1}$ and $t_n = 2M_n$. Also assume that

(7) $$\limsup \mu_n < \infty \quad ,$$

(8) $$\limsup \gamma_n < \infty \quad .$$

Then

(9) $$n^{1/2}\,\sigma_{n1}^{-1}\,\{\Delta_1(\hat{F},M_n) - \Delta_1(F_n,M_n)\} = n^{1/2}\,\sigma_{n1}^{-1}\int_0^{M_n} (\hat{F} - F_n)\, h_{n1} + o_p(1)$$
$$\to_d N(0,1) \ ,$$

where σ_{n1} is the σ_n of (5) with h_n replaced by h_{n1}.

(b) Assume that $\{F_n\}$ in $H_0 \cup H_1$ have densities $\{f_n\}$ and set

$$h_{n2}(x) = [0 < x < M_n] \int_0^x f_n(x-t)\, f_n(t)\, dt$$

$$+ [M_n \le x < 2M_n] \int_{M_n - x}^{M_n} f_n(x-t)\, f_n(t) dt - 2 \int_0^{M_n} f_n(x+t)\, f_n(t) dt \; .$$

Assume that (C1) and (C2) are satisfied by $\{F_n\}$, $G_1,\ldots,G_n$, and h_n replaced by h_{n2}. Then

$$(10) \qquad n^{1/2}\,\sigma_{n2}^{-1}\,\{\Delta_2(\hat{F},M_n) - \Delta_2(F_n,M_n)\} = n^{1/2}\,\sigma_{n2}^{-1} \int_0^{M_n} (\hat{F} - F_n)\, h_{n2} + o_p(1)$$

$$\to_d N(0,1) \quad ,$$

where σ_{n2} is the σ_n of (5) with $h_n = h_{n2}$.

Outline of Proof: Due to the limited space, we only sketch a proof of (9). The details for (10) are similar in nature. Write M for M_n and observe that

$$n^{1/2}(\Delta_1(\hat{F},M) - \Delta_1(F_n,M)) = n^{1/2} \int_0^M \int_0^M \{\hat{F}(s+t) - F_n(s+t)\}\, ds\, dt$$

$$- n^{1/2}\,\{(\int_0^M \hat{F})^2 - (\int_0^M F_n)^2\} = A_n - B_n, \text{ say} \quad .$$

One can check, using (7), that if (C1) and (C2) hold with $h_n = h_{n1}$, then they also hold with $h_n = 1$. Therefore, by (6)

$$\int_0^{M_n} (\hat{F} - F_n) = o_p(1) \quad .$$

This, in turn, implies that

(11) $$B_n = n^{1/2} \int_0^M (\hat{F} - F_n)\cdot\{ \int_0^M (\hat{F} + F_n)\} = 2 \int_0^M F_n \cdot n^{1/2} \int_0^M (\hat{F} - F_n) + o_p(1)\cdot$$

$$= 2\, \mu_n\, n^{1/2} \int_0^M (\hat{F} - F_n) + o_p(1) \quad .$$

Direct integration yields that

(12) $$A_n = n^{1/2} \int_0^{2M} \{u[0 < u < M] + (2M - u)\ [M \leq u < 2M]\}\ (\hat{F}(u) - F_n(u))\, du \quad .$$

Therefore, (12) and (11) yield the equality of (9), whereas the convergence in distribution to N(0,1) follows from the Main Theorem.

REMARK 1. Under H_0, $\mu_n = \lambda^{-1}$ and $h_{n1}(x) \to (x - 2/\lambda)$.

Actually, if

(13) $$\limsup \int_0^\infty x^2\, e^{-x}\, \{\bar{G}(x/\lambda)\}^{-1}\, dx < \infty \ ,$$

then one can show that

(14) $$\sigma_{n1}^2 = \lambda^{-4} \int_0^\infty (x - 1)^2\, e^{-x}\, \{\bar{G}(x/\lambda)\}^{-1}\, dx + o(1) \ .$$

Also, if

(15) $$\limsup \int_0^\infty y^2\, e^{-3y}\, \{\bar{G}(y/\lambda)\}^{-1}\, dy < \infty \ ,$$

then

(16) $$\sigma_{n2}^2 = 4^{-1} \int_0^\infty (y - 1/2)^2 e^{-3y} \{\bar{G}(y/\lambda)\}^{-1} dy + o(1) .$$

Thus (13) and (15) imply (C2), respectively, for h_{n1} and h_{n2} under H_0. A sufficient condition for (C1) to hold for both, h_{n1} and h_{n2}, under H_0 is that

(17) $$n^{-1/2} (\ell n\, n)^2 M_n^2 e^{3M_n\lambda} [\{\bar{G}(M_n)\}^{-3} - 1] = o(1) .$$

From (14) and (16) it is clear that the asymptotic null distribution of the proposed tests depend on λ and $\bar{G}$. To implement the tests we estimate λ by

$$\hat{\lambda} = \sum\delta_i / \sum Z_i$$

and $\bar{G}$ by

$$\hat{\bar{G}}(t) = \prod_{j=1}^{n} \left\{\frac{1+N(Z_j)}{2+N(Z_j)}\right\}^{[\delta_j = 0,\ Z_j \le t]} , \quad t \ge 0 .$$

It is easy to check that $\hat{\lambda}$ is a consistent estimator of λ under H_0 as long as

$$0 < \lim\inf \lambda \int_0^\infty \bar{G}(t)\, e^{-\lambda t} dt \le \lim\sup \lambda \int \bar{G}(t)\, e^{-\lambda t} dt < 1 .$$

That $\hat{\bar{G}}$ is a consistent estimator of $\bar{G}$, under H_0, can be deduced from Koul, Susarla and Van Ryzin (1981).

Let

$$\hat{\sigma}_{n1}^2 = \hat{\lambda}^{-4} \int_0^{N_n} (x - 1)^2 e^{-x} \{\hat{\bar{G}}(x/\hat{\lambda})\}^{-1} dx ,$$

$$\hat{\sigma}_{n2}^2 = 4^{-1} \int_0^{N_n} (y - 1/2)^2 e^{-3y} \{\hat{\bar{G}}(y/\hat{\lambda})\}^{-1} dy ,$$

where $N_n \uparrow \infty$.

Under (13), (15) and (17), with M_n replaced by N_n, one can show that $\hat{\sigma}_{nj}^2 = \sigma_{nj}^2 + o_p(1)$, $j = 1,2$. Consequently the test that rejects H_0 when $\Delta_j(\hat{F},M_n) \le z_\delta \hat{\sigma}_{nj}/n^{1/2}$ has the asymptotic size δ, $j = 1,2$. Next, consider

(b) H_0 vs H_2. Two reasonable measures of the deviation of H_2 from H_0, for a given F, are

$$\Delta_3(F) = \int\int [0 < s \le t < \infty]\ E(s,t)\ ds\,dt$$

and

$$\Delta_4(F) = \int\int [0 < s \le t < \infty]\ E(s,t)\ dF(s)\ dF(t) = \int (3F^2 - F - 2F^4)/6 ,$$

where

$$E(s,t) = F(t)\ J(s) - F(s)\ J(t),\ 0 < s \le t < \infty .$$

Let $\Delta_3(F,M) = \int\int [0 < s \le t \le M]\ E(s,t) ds\,dt$ and define $\Delta_4(F,M)$, similarly. The test j rejects H_0 in favor of H_2 if $\Delta_j(\hat{F},M_n)$ is large, $j = 3,4$. The following theorem gives the asymptotic normality of these test statistics for a sequence $\{F_n\}$ in $H_0 \cup H_2$ and for non-identically distributed censoring variables. Note that a variant of the Δ_4-test was suggested by Chen, Hollander and Langberg (1980) but they do not discuss the asymptotic distribution under sequences of alternatives.

THEOREM 3:

(a) Let $\{F_n\}$ be in $H_0 \cup H_2$. For $s \le M_n$,

$$h_{n3}(s) := 2\int_0^{M_n} \min(s,x)\ F_n(x)dx - \int_0^{M_n} (x - s)\ F_n(x)dx - \int_0^s (s-x)F_n(x)dx.$$

Assume $\{F_n\}$, $G_1,\ldots,G_n$ and h_{n3} satisfy (C1) and (C2). Also, assume that (8) holds.

Then

$$n^{1/2}\,\sigma_{n3}^{-1}\,(\Delta_3(\hat{F},M_n) - \Delta_3(F_n,M_n)) = n^{1/2}\,\sigma_{n3}^{-1}\int_0^{M_n} (\hat{F} - F_n)\,h_{n3} + o_p(1)$$

$$\to_d N(0,1)\quad ,$$

where σ_{n3} is the σ_n of (5) with $h_n = h_{n3}$.

(b) Let

$$h_{n4} := (6F_n - 1 - 8F_n^{\,3})/6 \text{ on } [0,M_n]\ .$$

Assume $\{F_n\}$, h_{n4} and $\{G_i\}$ satisfy (C1) and (C2). Also, assume that (7) holds. Then

$$n^{1/2}\,\sigma_{n4}^{-1}\,(\Delta_4(\hat{F},M_n) - \Delta_4(F_n,M_n)) = n^{1/2}\,\sigma_{n4}^{-1}\int_0^{M_n} (\hat{F} - F_n)\,h_{n4} + o_p(1)$$

$$\to_d N(0,1)\quad ,$$

where σ_{n4} is the σ_n of (5) with $h_n = h_{n4}$.

REMARK 2. As in Remark 1, it can be shown that under (13) and under H_0 with a fixed λ,

$$\sigma_{n3}^{\,2} = \lambda^{-6}\int_0^{\infty} e^{-t}\,(2 - 2e^{-t} - t)^2\,\{\bar{G}(t/\lambda)\}^{-1}\,dt + o(1)$$

and that if

$$\lim\sup \int_0^{\infty} e^{-\lambda}\,\{\bar{G}(s/\lambda)\}^{-1}\,ds < \infty\ ,$$

then

$$\sigma_{n4}^{\,2} = (36\lambda^2)^{-1}\int_0^{\infty} e^{-s}\,(3e^{-s} - 1 - 2e^{-3s})^2\,\{\bar{G}(s)/\lambda)\}^{-1}\,ds + o(1)\ .$$

Moreover, a sufficient condition for (C1) to hold under H_0, for both h_{n3} and h_{n4}, is (17).

Let

$$\hat{\sigma}_{n3}^2 = (\hat{\lambda})^{-6} \int_0^{N_n} e^{-t} (2 - 2e^{-t} - t)^2 \{\hat{\bar{G}}(t/\hat{\lambda})\}^{-1} dt \quad ,$$

$$\hat{\sigma}_{n4}^2 = (36\hat{\lambda}^2)^{-1} \int_0^{N_n} e^{-t} (3e^{-t} - 1 - ze^{-3t})^2 \{\hat{\bar{G}}(t/\hat{\lambda})\}^{-1} dt \ .$$

Then the test that rejects H_0 when $\Delta_j(\hat{F},M_n) \geq z_{1-\alpha}\, \hat{\sigma}_{nj}/n^{1/2}$ has the asymptotic size α, $j = 3,4$. Both of these tests are consistent against a fixed F in H_2 and for all those censoring distributions for which (17) holds and an analogue of (C2) holds for h_{n3} and h_{n4} at the given F.

3. Asymptotic Efficiency

Consider the problem of testing H_0 vs a sequence of alternatives $\{F_{\theta_n}\} \in H_1$ when there is no censoring. In this case one can base tests on $\tilde{\Delta}_j = \Delta_j(\tilde{F})$, $j = 1,2$, where $\tilde{F}(x) = n^{-1} \sum[X_i > x]$, $x \geq 0$. Observe that $\tilde{\Delta}_1 = (2n)^{-1} \sum X_i^2 - \bar{X}^2$. The $\tilde{\Delta}_2$ is a priori scale invariant while a scale invariant analogue of $\tilde{\Delta}_1$ is

$$\Delta_1^* = \tilde{\Delta}_1 / \bar{X}^2 \quad .$$

Note that an analogue of Δ_1^* under random censoring is $\hat{\lambda}^2 \Delta_1(\hat{F},M_n)$. We did not consider this statistic in the previous section because its asymptotic null distribution still depends on λ as does that of $\Delta_1(\hat{F},M_n)/(\int_0^{M_n} \hat{F})^2$.

Using the standard central limit theorem one has

$$(18) \qquad n^{1/2} (\Delta_1^* - \mu_n^{-2} \Delta_1(F_{\theta_n})) \to N(0,1) \ , \ (\mu_n = \int_0^\infty F_{\theta_n}) \ ,$$

for all $\{F_{\theta_n}\} \varepsilon H_1$ which are contiguous to F_{θ_0}. Note that implicit in (18) is the assumption that $\Delta_1(F_{\theta_n}) < \infty$ for all n which amounts to assuming the finiteness of the second moments (see (8)) whereas no such assumption is needed for $\tilde{\Delta}_2$-test.

Now if $\dot{\Delta}_j(\theta) = \partial\Delta_j(F_\theta)/\partial\theta$, $j = 1,2$, then it follows that the asymptotic relative Pitman efficiency of Δ_1^*-test relative to the $\tilde{\Delta}_2$-test is

$$e(1,2) = \frac{5}{432} \{\dot{\Delta}_1(\theta_0)/\dot{\Delta}_2(\theta_0)\}^2 .$$

Consider the alternatives: (a1). $F_{\theta_n}(x) = e^{-x-x^2\theta_n/2}$, $\theta_n = \delta_n^{-1/2}$, $\delta > 0$, $x \geq 0$. Then $\theta_0 = 0$ and $\dot{\Delta}_1(0) = 1$, $\dot{\Delta}_2(0) = 1/16$ and $e(1,2) = (5x256)/432 = 2.96$.

(a2). If $F_{\theta_n}(x) = \exp(-x^{\theta_n})$, $\theta_n = 1 + \delta n^{-1/2}$, $\delta \geq 0$, then $\theta_0 = 1$ and $\dot{\Delta}_1(0) = 1$, $\dot{\Delta}_2(1) = 1/8$ and $e(1,2) = .74$.

Now suppose there is random censoring with $\bar{G}(x) \equiv e^{-\theta x}$, $\theta < \lambda$. Then from (14)

$$\sigma_{n1}^2 \to \lambda^{-4} \int (x-1)^2 e^{-\alpha x} dx \qquad (\alpha = 1-(\theta/\lambda))$$

$$= \lambda^{-4} [2\alpha^{-3} - 2\alpha^{-2} + \alpha^{-1}]$$

$$= (1+r^2)/\lambda^4\alpha^3 = \sigma_1^2, \text{ (say)} . \qquad (r = \theta/\lambda)$$

Also, from (16)

$$\sigma_{n2}^2 \to 4^{-1} \int (x - 1/2)^2 e^{-(3-r)x} dx \qquad (\beta = 3-r)$$

$$= 4^{-1} [2\beta^{-3} - \beta^{-2} + 4^{-1}\beta^{-1}]$$

$$= (5 - 2r+r^2)/16\beta^3 = \sigma_2^2, \text{ (say)} .$$

Note that $\dot{\Delta}_j$'s do not change. Then at the alternative (a1),

$$e(1,2) = 256.\ (\sigma_2^2/\sigma_1^2) = 256.\ \frac{(5-2r+r^2)\ \lambda^4 (1-r)^3}{16(3-r)^3\ (1+r^2)}$$

$$\to 2.96 \qquad \text{as } r \to 0 \quad (\text{i.e., } \theta \to 0)$$

$$\to 0 \qquad \text{as } r \to 1 \quad (\text{i.e., } \theta \to \lambda) \ .$$

Thus, for example, if censoring distributions are almost like the exponential (λ) distributions, then Δ_2-test would be preferred.

In general, if $\bar{G}$, the average of censoring distributions, has lighter right tail compared to the exponential tails, we suggest using the test based on $\Delta_1(\hat{F}, M_n)$ with $M_n = c(\ln n)^a$, $c > 0$, $0 < a < 1$.

4. Proof of the Main Theorem

The technical details of the proof are similar to those in Section 7 of Koul, Susarla and Van Ryzin (1981). We provide only a sketch of the proof here. Write $\hat{F} = \hat{H}\hat{W}$ where $(n+1)\ \hat{H} = 1+N$ and $\hat{W} = \hat{\bar{G}}^{-1}$, the second factor in (4). Write M for M_n. Observe that

$$\hat{F} - F_n = \bar{G}^{-1}\ (\hat{H} - H) + \hat{H}\ (\hat{W} - \bar{G}^{-1}), \quad H = \bar{G}F_n \ .$$

Hence,

$$n^{1/2} \int_0^M (\hat{F} - F_n)h_n = n^{1/2} \left[\int_0^M \hat{H}(\hat{W} - \bar{G}^{-1})h_n + \int_0^M \bar{G}^{-1}\ (\hat{H} - H)h_n\right]$$

$$= I + II, \ (\text{say}) \ .$$

The term II is a sum of centered independent r.v's. We only need to approximate I by a sum of independent r.v's. To this end we write $W = \exp(\ln W)$, $\bar{G}^{-1} = \exp(-\ln \bar{G})$, and use a Taylor expansion to obtain

$$(19)\qquad |(\hat{W}-\bar{G}^{-1}) - \bar{G}^{-1}(\ln\hat{W} + \ln\bar{G})| \leq 2\bar{G}^{-1}(\ln\hat{W} + \ln\bar{G})^2 .$$

From the details similar to those in Section 7 of Koul, Susarla and Van Ryzin (use Lemma 7.1 with $p_i = F_n G_i$ and the details similar to those in the proof of Lemma 7.2), one obtains

$$E(\text{RHS } (19)) \leq -k_1\, n^{-1}\,\bar{G}^{-1}\int_0^{\cdot} F_n\, H^{-4}\, d\bar{G}, \quad (\text{for some constant } k_1) .$$

Therefore,

$$I = n^{1/2}\int_0^M h_n\,\hat{H}\,\bar{G}^{-1}(\ln\hat{W} + \ln\bar{G}) + o_p(1).$$

provided $n^{1/2}\int_0^M |h_n(x)|\bar{G}^{-1}(x)(\int_0^x F_n\, H^{-4}\, d\bar{G})dx = o(1)$, which in turn is implied by (C1). The next step is to approximate $\ln\hat{W}$. Again, carrying out details similar to those in Koul, Susarla and Van Ryzin, one obtains

$$(20)\qquad I = \int_0^M F_n h_n\, n^{1/2}\{\int_0^x (2H - H_n)H^{-2}\, dH_n^* + \ln\bar{G}(x)\} + o_p(1) ,$$

where

$$H_n = n^{-1}N,\ nH_n^*(\cdot) = \sum(1-\delta_i)\,[Z_i \leq \cdot] .$$

The first r.v. on the right-hand side of (20) can be expressed as a U-statistic and, hence, by the projection technique, one can show that, under (C1),

$$(21)\qquad I = \int_0^M h_n(x)\, F_n(x)\, n^{-1/2}\sum_i \{[(1-\delta_i)][Z_i \leq x]\, H^{-1}(Z_i) - \int^{x\wedge Z_i} H^{-2} d\bar{G}$$

$$+ \ln\bar{G}(x)\}dx + o_p(1) .$$

Combining (21) with (19), the final approximating r.v. is the sum of II and the first r.v. on the right hand side of (21). Its variance is σ_n^2 and (C2) implies the asymptotic normality (6) by the Lindeberg-Feller CLT.

ACKNOWLEDGEMENT

The research of author (VS) is supported by the National Institutes of Health Grant No. 1-RO-1-GM-28405. This author is on leave from Suny, Binghamton, New York.

REFERENCES

Barlow, R.E. and Proschan, F. (1975). Statistical Theory of Reliability and Life Testing. Holt, Rinehart and Winston, Inc.

Chen, Y.Y., Hollander, M. and Langberg, N. (1980). Tests for monotone mean residual life using randomly censored data. Statistics Report, No:M459, Florida State University.

Bryson, M.C. (1974). Heavy tailed distributions: properties and tests. Technometrics **16**, 61-68.

Bryson, M.C. and Siddiqui, M.M. (1969). Some criteria for aging. Journal of the American Statistical Association **64**, 1472-1483.

Hollander, M. and Proschan, F. (1972). Testing whether new is better than used. Annals of Mathematical Statistics **43**, 1136-1146.

Hollander, M. and Proschan, F. (1975). Tests for mean residual life. Biometrika **62**, 585-593.

Koul, H.L. (1977). A test for new is better than used. Communications in Statistics, Theory and Methods **A6**, 563-573.

Koul, H.L. (1978a). A class of tests for testing 'new is better than used'. Canadian Journal of Statistics **6**, 249-271.

Koul, H.L. (1978b). Testing for new is better than used in expectation. Communications in Statistics, Theory and Methods **A7**, 685-701.

Koul, H.L. and Susarla, V. (1980). Testing for new better than used in expectation with incomplete data. Journal of the American Statistical Association 75, 952-956.

Koul, H.L., Susarla, V. and Van Ryzin, J. (1981). Regression analysis with randomly right censored data. Annals of Statistics 9, 1276-1288.

Marshall, A.W. and Proschan, F. (1972). Classes of distributions applicable in replacement, with renewal theory implications. Proceedings of the Sixth Berkeley Symposium on Mathematical Statistics and Probability, Vol.I, University of California Press, Berkeley, California, 395-415.

NEGATIVE DEPENDENCE

Henry W. Block, Thomas H. Savits

Department of Mathematics and Statistics,
University of Pittsburgh

Moshe Shaked

Department of Mathematics, University of Arizona

1. Introduction

Various concepts of positive dependence have been considered in the literature. Many of these concepts have been developed to obtain conditions on a random vector $\underline{T} = (T_1,\dots,T_n)$ such that

$$P\{T_1 > t_1,\dots,T_n > t_n\} \geq \prod_{i=1}^{n} P\{T_i > t_i\} \text{ for all } t_1,\dots,t_n \quad . \tag{1}$$

See Lehmann (1966) for a discussion of the bivariate case and Barlow and Proschan (1975) and Block and Ting (1981) for details in the multivariate case.

There are many distributions such as the multinomial and the Dirichlet for which the reverse inequality holds (see Mallows (1968) and Jogdeo and Patil (1975)). However no systematic study of negative dependence concepts in the $n \geq 3$ case was attempted until recently (the bivariate case was considered by Lehmann (1966)).

In this paper we discuss various multivariate concepts of negative dependence. Many of these concepts arose out of discussions which were begun at the NSF/CBMS conference held in Columbia, Missouri in June 1979 at which Frank

Proschan was the principal speaker. Sections 2 and 3 contain results concerning positive dependence. Results on the positive dependence of the multivariate normal are given in Section 4. Section 5 deals with some basic negative dependence results in the bivariate case. Various concepts of multivariate negative dependence are compared in Section 6. These are based on topics considered in Block, Savits and Shaked (1982), Ebrahimi and Ghosh (1981) and Karlin and Rinott (1980). One of the fundamental results discussed is that if a distribution satisfies an intuitive structural condition called Condition N, then it satisfies all of the other conditions mentioned including the fundamental negative dependence inequality given by (3). Condition N is satisfied by the multinomial, hypergeometric and Dirichlet distribution as well as several others. Section 7 concludes with two other negative dependence conditions. One is a concept of negative association introduced by Jogdeo and Proschan (1981) and a second deals with a condition of stochastic ordering due to Block, Savits and Shaked (1981).

2. Positive Dependence - Bivariate Case

For the bivariate random vector (S,T) where $\mathrm{cov}(S,T) \geq 0$ we often have or want to show that

$$P\{S > s,\ T > t\} \geq P\{S > s\}\, P\{T > t\} \text{ for all } s,t \quad . \tag{2}$$

If this inequality holds we have a lower bound for the joint survival function in terms of the marginal survival functions. We say that (S,T) is positive quadrant dependent (PQD) if (2) holds. Condition (2) is often difficult to verify directly for specific bivariate distributions, so that a condition which implies (2) and is easy to check is useful. One such condition is that (S,T) has a joint density f(s,t) which is TP_2, i.e.,

$$f(s,t)\, f(s',t') \geq f(s,t')\, f(s',t) \text{ for all } t < t',\ s < s' \quad .$$

This has been also called positive likelihood ratio dependent by Lehmann (1966) who establishes the following relationships. He shows that if (S,T) has a TP_2 density then $P\{T > t | S = s\}$ increases in s (called positive regression dependence). This last condition implies that (S,T) is PQD which in turn implies that $\text{cov}(S,T) \geq 0$.

3. Positive Dependence - Multivariate Case

A parallel multivariate theory has been developed (see Barlow and Proschan (1975)). The various notions are:

1) $\underline{T} = (T_1,\ldots,T_n)$ has a joint density which is TP_2 (totally positive of order 2) in pairs, i.e., the joint density $f(t_1,\ldots,t_n)$ is TP_2 in t_i and t_j for any $i \neq j$ in $\{1,2,\ldots,n\}$ when the remaining variables are fixed (there are also some conditions on the support of f, see Kempermann (1977));

2) $P\{T_i > t | T_1 = t_1,\ldots,T_{i-1} = t_{i-1}\}$ increases in $t_1,\ldots,t_{i-1}$ for $i = 2,\ldots,n$ and for all t;

3) $\underline{T}$ is associated which means that $\text{cov}(f(\underline{T}), g(\underline{T})) \geq 0$ for all functions f and g which increase componentwise;

4) $P\{T_1 > t_1,\ldots,T_n > t_n\} \geq \prod_{i=1}^{n} P\{T_i > t_i\}$ for all $t_1,t_2,\ldots,t_n$;

5) $\text{cov}(T_i,T_j) \geq 0$ for all $i \neq j$ in $\{1,2,\ldots,n\}$.

Each of these conditions implies the following condition. The most important result is that 1) implies 4). For recent developments concerning multivariate positive dependence conditions see Block and Ting (1981). A new concept of positive dependence by stochastic ordering has been given by Block, Savits and Shaked (1981).

4. Positive Dependence of the Multivariate Normal

The results of the previous section have been applied to the multivariate normal distribution. Conditions on the covariance matrix have been given so that the various positive dependence conditions are satisfied. These are summarized below.

Let $\underline{X} = (X_1,\ldots,X_n)$ be a random vector having a multivariate normal distribution with mean vector $\underline{0}$ and covariance matrix $\underline{\Sigma}$. The strongest condition, TP_2 in pairs, is satisfied if and only if $\underline{\Lambda} = \underline{\Sigma}^{-1}$ exists and has nonpositive off diagonal elements. A weaker condition is that $\underline{\Sigma}$ has nonnegative elements and this is necessary and sufficient for $\underline{X}$ to be associated as was recently shown by Pitt (1982) (see also Jogdeo and Perlman (1981)). Since condition 4) implies condition 5), it is also clear, because of the previous result, that the nonnegativity of the elements of $\underline{\Sigma}$ is also necessary and sufficient for 4) to hold.

Summarizing, for the multivariate normal the main result is that if $\mathrm{cov}(X_i,X_j) \geq 0$ for all i,j in $\{1,\ldots,n\}$ then

$$P\{X_1 > x_1,\ldots,X_n > x_n\} \geq \prod_{i=1}^{n} P\{X_i > x_i\} \text{ for all } x_1,\ldots,x_n \quad .$$

This result is not true for most multivariate distributions. For related results concerning the dependence of $(|X_1|,|X_2|,\ldots,|X_n|)$ and the comparison of two normal vectors see Block and Sampson (1982).

5. Negative Dependence - Bivariate Case

The results of this section are practically identical to those of the bivariate positive dependence case and can be obtained by reversing inequalities and monotonicities (or by considering (S,-T) where (S,T) satisfies one of the positive dependence conditions). See Lehmann (1966) for details.

The negative dependence analog of TP_2 is RR_2 (or reverse rule of order 2). The vector (S,T) is said to be RR_2 if (S,T) has density f(s,t) which satisfies

$$f(s,t)\, f(s',t') \leq f(s,t')\, f(s',t) \text{ for all } t < t',\ s < s' \ .$$

This condition implies that $P\{T > t | S = s\}$ decreases in s for all t. The latter condition implies that $P\{S > s, T > t\} \leq P\{S > s\}\, P\{T > t\}$ for all s and t from which it follows that $\mathrm{cov}(S,T) \leq 0$.

6. Negative Dependence - Multivariate Case

Until recently negative dependence concepts in the $n \geq 3$ case had not been widely studied. One reason for this was that the structure of negative dependence was not generally understood. Recent studies have shown that negative dependence is quite different from positive dependence.

Ebrahimi and Ghosh (1981) have reversed inequalities and directions of monotonicities for the multivariate positive dependence concepts and compared resulting concepts. These analogs of the positive dependence concepts do not in general imply the condition

$$(3) \qquad P\{T_1 > t_1, \ldots, T_n > t_n\} \leq \prod_{i=1}^{n} P\{T_i > t_i\} \text{ for all } t_1, \ldots, t_n \ .$$

The condition that $P\{T_i > t_i | T_1 > t_1, \ldots, T_{i-1} > t_{i-1}\}$ is decreasing in $t_1, \ldots, t_{i-1}$ for all t_i, $i = 2, \ldots, n$ does imply (3), but it is no easier to check than (3) itself.

We now illustrate some of the problems in obtaining a condition which implies (3). Since TP_2 in pairs implies (1), it might seem reasonable to hope that the negative dependence analog given by Definition 1 below would imply (3).

Definition 1: The random vector $\underline{T} = (T_1, \ldots, T_n)$ is said to be RR_2 in pairs if it has joint density $f(t_1, \ldots, t_n)$ which is RR_2 in any two variables when the others are fixed.

This definition does not imply (3). Moreover not even the marginals of $\underline{T}$ need to be RR_2 in pairs. A modified defintion was given by Ebrahimi and Ghosh (1981).

Definition 2: The random vector $\underline{T} = (T_1, \ldots, T_n)$ is said to be completely RR_2 in pairs if the joint density and all of its marginals are RR_2 in pairs.

It remains an unsolved problem as to whether this implies (3). A further attempt to modify this definition was made by Block, Savits and Shaked (1982). This still stronger definition follows.

Definition 3: The random vector $\underline{T} = (T_1,\ldots,T_n)$ has an RR_2 in pairs measure if $\mu(I_1,I_2,\ldots,I_n) = \int_{I_1}\cdots\int_{I_n} d\,P\{T_1 \le t_1,\ldots,T_n \le t_n\}$ is RR_2 in pairs, e.g., $\mu(I_1,I_2,\ldots,I_n)$ is RR_2 in I_1 and I_2 if

$$\mu(I_1,I_2,I_3,\ldots,I_n)\,\mu(I_1',I_2',I_3,\ldots,I_n)$$
$$\le \mu(I_1,I_2',I_3,\ldots,I_n)\,\mu(I_1',I_2,I_3,\ldots,I_n)$$

for all intervals $I_1 < I_1'$ and $I_2 < I_2'$ where $I_i < I_i'$ means every point in I_i is less than every point in I_i'.

Block, Savits and Shaked (1982) prove that if $\underline{T}$ satisfies Definition 3 then $\underline{T}$ satisfies (3). The only problem is that Definition 3 is not easy to check. In the same paper these authors propose a structural condition, which many standard multivariate distributions satisfy. This is Condition N below.

Condition N: The random vector $\underline{T} = (T_1,\ldots,T_n)$ is such that there exist independent r.v.'s $S_0,S_1,\ldots,S_n$ having PF_2 densities (i.e., the densities $f_i(s_i)$, $i = 0,\ldots,n$ satisfy $f_i(t_1 - t_2)$ is TP_2 in (t_1,t_2)) and there exists a real number s such that

$$(T_1,\ldots,T_n) \overset{st}{=} [(S_1,\ldots,S_n) \mid S_0 + S_1 + \cdots + S_n = s] .$$

Note - For a discussion of PF_2 (Polya frequency of order 2) densities see Barlow and Proschan (1975). Densities with these properties have increasing failure rates.

Essentially a random vector $\underline{T}$ satisfying this condition is like the multinomial distribution, i.e., the sum of the components of the random vector is fixed. Remarkably it turns out that many of the standard multivariate distributions which have negatively correlated univariate marginals satisfy this condition. It can be shown that the multinomial, the symmetric normal with negative correlations, the multivariate hypergeometric and the Dirichlet all satisfy this definition.

Example: The Dirichlet distribution has the same distribution as the conditional distribution of a sum of certain independent gamma distributions given that their sum is equal to 1 and it is easy to show that those gammas have PF_2 densities. Thus, the Dirichlet satisfies Condition N.

Furthermore, Condition N can be seen to imply Definition 3 and so (3) is satisfied. Thus Condition N is an intuitive condition, which is easy to check, and which implies the inequality (3).

Karlin and Rinott (1980) have also proposed a definition which implies the basic inequality (3). Their definition is also a strengthening of the RR_2 condition.

Definition 4: The random vector $\underline{T} = (T_1,\ldots,T_n)$ having joint density f is called strongly multivariate RR_2 (S-MRR_2) if for any set of PF_2 functions $\phi_1,\ldots,\phi_{n-k}$, $2 \leq k \leq n$ the function

$$g(x_{v_1},\ldots,x_{v_k}) = \int\cdots\int f(x_1,\ldots,x_n)\,\phi_1(x_{j_1})\ldots\phi_{n-k}(x_{j_{n-k}})\,dx_{j_1},\ldots,dx_{j_{n-k}}$$

is RR_2 in pairs of the unintegrated variables $x_{v_1},\ldots,x_{v_k}$.

Note - The case $k = n$ in the above definition corresponds to assuming the density is strictly positive and RR_2 in pairs on a product set in R^n.

As mentioned above if $\underline{T}$ is S-MRR_2 then $\underline{T}$ satisfies (3), but as can be seen from the computations in Karlin and Rinott (1980) this definition is cumbersome to check. Fortunately, if $\underline{T}$ satisfies Condition N, then it is S-MRR_2 (see Block, Savits and Shaked (1982)).

Summarizing the relationships, Condition N is stronger than Definition i which is stronger than Definition $i - 1$ for $i = 2,3,4$ and Definition 3 implies (3).

7. Other Negative Dependence Conditions

One of the problems of negative dependence was that there did not seem to be a natural analog of the positive dependence concept of association. Simply reversing inequalities or monotonicities in the definition of association leads to concepts which have certain anomalies. For example, assume

$$\text{cov}(f(\underline{T}), g(\underline{T})) \leq 0 \text{ for all increasing } f,g .$$

Then for f and g which are functions only of the ith variables it must be that $\text{cov}(f(\underline{T}), g(\underline{T})) = 0$. A similar circumstance occurs when f and g are taken to be the same function.

Jogdeo and Proschan (1981) and Alam and Saxena (1981) have proposed a definition which avoids these difficulties. They define $\underline{T} = (T_1,\ldots,T_n)$ to be negatively associated if

$$\text{cov}(f(T_i,\ i \in \Lambda),\ g(T_i,\ i \in \bar{\Lambda})) \leq 0$$

where $\Lambda \subset \{1,2,\ldots,n\}$ and $\bar{\Lambda} = \{1,2,\ldots,n\}\setminus\Lambda$ and f and g are any nondecreasing functions of the appropriate number of variables.

Jogdeo and Proschan (1981) have shown that if $\underline{T}$ is negatively associated then random variables defined as increasing functions on disjoint subsets of $\underline{T}$ are negatively associated. The negative association condition can be seen to imply (3), but the condition is not easy to check. However, these authors have essentially shown that negative association is implied by Condition N. Thus, all of the distributions which satisfy Condition N are negatively associated.

One other condition has been proposed by Block, Savits and Shaked (1981). One of the motivations was to obtain a condition satisfied by a wider class of multivariate normals than satisfies Condition N. Although Condition N is a natural condition for distributions like the multinomial and the Dirichlet it is not natural for multivariate normal distributions. The new condition resembles, but is more general than, a condition used by Mallows (1968) and by Jogdeo and Patil (1975) to show (3) for specific distributions. The condition used by these authors was $P\{X_2 > x_2,\ldots,X_n > x_n | X_1 > x_1\}$ is decreasing in x_1 for all $x_2,\ldots,x_n$. The new condition called negatively dependent by stochastic ordering is that $E(f(X_1,\ldots,X_{i-1}, X_{i+1},\ldots,X_n) | X_i = x_i)$ is decreasing in x_i for all nondecreasing functions f and all $i = 1,\ldots,n$.

Several natural models and distributions which satisfy this condition are:

1) Condition N (actually a slightly stronger version);

2) $\underline{T} = (T_1,\ldots,T_n)$ such that there exists independent, identically distributed, and continuous $X_1,\ldots,X_n$ and a real number z such that

$$\underline{T} \overset{st}{=} [(X_1,\ldots,X_n) \mid \min(X_1,\ldots,X_n) = z] ;$$

3) $\underline{T}$ is such that there exist independent and identically normally distributed $X_1,\ldots,X_n$ such that

$$\underline{T} \overset{st}{=} (X_1 - \bar{X},\ldots,X_n - \bar{X}) ;$$

4) $\underline{T}$ is multivariate normal with nonpositive correlations;

5) All of the distributions mentioned in Section 6.

ACKNOWLEDGEMENTS

Drs. Block and Savits are supported by the U.S. Office of Naval Research and Dr. Shaked by the National Science Foundation.

REFERENCES

Alam, K. and Lal Saxena, K.M. (1981). Positive dependence in multivariate distributions. Communications in Statistics A(10), 1183-1196.

Barlow, R.E. and Proschan, F. (1975). Statistical Theory of Reliability and Life Testing. Holt, Rinehart and Winston, New York.

Block, H.W. and Sampson, A. (1982). Inequalities on distributions: bivariate and multivariate. In Encyclopedia of Statistical Sciences, Vol. 10 (Eds. N.L. Johnson and S. Kotz), To appear.

Block, H.W., Savits, T.H. and Shaked, M. (1981). A concept of negative dependence using stochastic ordering. Unpublished report.

Block, H.W. Savits, T.H. and Shaked, M. (1982). Some concepts of negative dependence. Annals of Probability, 10, 773-779.

Block, H.W. and Ting, M.-L. (1981). Some concepts of multivariate dependence. Communications in Statistics, A(10), 749-762.

Ebrahimi, N. and Ghosh, M. (1981). Multivariate negative dependence. Communications in Statistics A(10), 307-337.

Jogdeo, K. and Patil, G.P. (1975). Probability inequalities for certain multivariate discrete distributions. Sankhyā, Series B, 37, 158-164.

Jogdeo, K. and Perlman, M. (1981). A simple proof establishing association for the positively correlated normal random variables. Unpublished report.

Jogedo, K. and Proschan, F. (1981). Negative association of random variables, with applications. Florida State University Statistics Report M-590.

Karlin, S. and Rinott, Y. (1980). Classes of orderings of measures and related correlation inequalities, II. Multivariate reverse rule distributions. Journal of Multivariate Analysis, 10, 499-516.

Kemperman, J.H.B. (1977). On the FKG-inequality measures on a partially ordered space. Proceedings of the Kon. Ned. Akad. Wet., Amsterdam, Series A, 80, 313-333.

Lehmann, R.L. (1966). Some concepts of dependence. Annals of Mathematical Statistics 43, 1137-1153.

Mallows, C.L. (1968). An inequality involving multinomial probabilities. Biometrika 55, 422-424.

Pitt, L. (1982). Positively correlated normal variables are associated. Annals of Probability 10, 496-499.

SOME RECENT RESULTS IN COMPETING RISKS THEORY

Asit P. Basu
University of Missouri

and

John P. Klein
Ohio State University

1. Introduction

The problems of competing risks and complementary risks arise quite naturally in a number of contexts, particularly in problems of survival analysis and reliability theory. The problems, in their simplest form, may be described as follows. Let X_i be a random variable with cumulative distribution function (C.D.F.) $F_i(x)$, $(i = 1,2,\ldots,p)$. We assume that the X_i's are not observable but $U = \min(X_1,\ldots,X_p)$ or $V = \max(X_1,\ldots,X_p)$ is. We would like to determine uniquely the marginal C.D.F.'s, F_i's, from that of U in the competing risks problem or from that of V in the complementary risks problem. We would also consider related inference problems.

As examples of the concepts consider the following:

(a) Let X_i be the time to death (failure) from cause C_i (of component C_i). Here X_i's are not observable but we observe a death time U (or time to series system failure) or a time V at which the last remaining duplicated organ fails (time to failure of a parallel system).

(b) In survival analysis randomly censored data correspond to the situation when $p = 2$, X_1 is the variable of interest and X_2 the censoring variable.

In this report we present results in three areas of competing risks analysis. In Section 2 the problem of identifiability is discussed. In Section 3 we look at dependent competing risks and at techniques for converting dependent models to independent models which preserve the models' properties, in a sense to be discussed later. In Section 4 accelerated life tests in a competing risks framework are considered.

2. Identifiability

2.1 Introduction

Before we can consider the inference problems, we need to resolve the question of identifiability. Basu (1981a,b) has given a survey of the identifiability problem in the parametric case. Consider the following definition of identifiability.

DEFINITION 1: Let U be an observable random variable with C.D.F. F_θ and let $F_\theta \in \mathcal{F} = \{F_\theta : \theta \in \Omega\}$, a family of distribution functions indexed by a parameter θ. Here θ could be scaler or vector valued. θ is said to be nonidentifiable by U if there is at least one pair (θ, θ'), $\theta \neq \theta'$, where θ and θ' both are in Ω, such that $F_\theta(u) = F_{\theta'}(u)$ for all u. In the contrary case we shall say θ is identifiable.

In many cases, where θ is not identifiable, there exists a non-constant function $\gamma(\theta)$ which is identifiable. That is, for any $\theta, \theta' \in \Omega$, $F_\theta(u) = F_{\theta'}(u)$ for all u implies $\gamma(\theta) = \gamma(\theta')$. In this case θ is said to be partially identifiable.

In case θ is not identifiable by U, it may be possible to introduce an additional random variable I so that θ is identifiable by the augmented random variable (U,I). In this case the original identifiability problem is called rectifiable.

EXAMPLE 1: Let X_i be independent random variables with density functions $f_i(x) = \lambda_i \exp(-\lambda_i x)$, $(i=1,2)$. Let $\theta = (\lambda_1, \lambda_2)$. Here θ is not identifiable by U. However, θ is partially identifiable since $\gamma(\theta) = \lambda_1 + \lambda_2$ is identifiable. The problem is also rectifiable since θ is identifiable by (U,I) where $I = i$ if $U = X_i$, $(i = 1,2)$.

Complementary risks is the dual of competing risks since $\max(X_1,\ldots,X_p) = -\min(-X_1,\ldots,-X_p)$. Usually it is sufficient to consider the results in terms of U. However, there are situations when V is analytically simpler to study.

2.2 Independent Random Variables

Assume $X_1,\ldots,X_p$ are independent but not identically distributed random variables. Let $I = k$ if $U = \min(X_1,\ldots,X_p) = X_k$. Let $S_i(x) = P(X_i > x)$ and $S_i^*(x) = P(U > x,\ I = i)$ $(i = 1,\ldots,p)$. Then Berman (1963) has proved the identifiability of the F_i's in the following theorem.

THEOREM 1: (Berman (1963))

$$S_k^*(x) = -\int_x^\infty [\prod_{\substack{j=1 \\ j\neq k}}^{p} S_j(t)]\, dS_k(t)$$

and

$$S_k(x) = \exp[\int_0^x (\sum_{j=1}^{p} S_j^*(t))^{-1}\, dS_k^*(t)] \quad .$$

Theorem 1 justifies the estimation of parameters in the regression problem of Miller (1976). For $p = 2$, Peterson (1977) extends the result to the case where S_1 and S_2 have no common jump points.

THEOREM 2: (Peterson (1977))

$$S_i(t) = \exp\left\{ \int \frac{dS_1^*(x)}{S_1^*(x) + S_2^*(x)} + \sum_{\substack{s:\text{Jump point of } S_i(\cdot) \\ s \leq t}} \ell n \left[\frac{S_1^*(s^+) + S_2^*(s^+)}{S_1^*(s^-) + S_2^*(s^-)} \right] \right\} .$$

Theorem 2 gives an alternate representation for the survival function and the Kaplan-Meier (1958) estimator. It also helps proving the strong consistency of the Kaplan-Meier estimator.

Next, consider the case of the non-identified minimum. Basu and Ghosh (1980) prove the following result.

THEOREM 3: (Basu and Ghosh (1980))

Let $\mathcal{F}$ be a family of probability density functions (p.d.f.) on R_1 with support on (a,b) which are continuous and are positive to the left of some point A and such that if f and g are any two distinct members of $\mathcal{F}$ then $\lim_{x \to a}\{f(x)/g(x)\}$ is either 0 or ∞. Let X_i be independent with p.d.f. f_i in $\mathcal{F}$ $(i = 1,2,\ldots,p)$ and Y_i be independent with p.d.f. $g_j \in \mathcal{F}$ $(j = 1,\ldots,q)$. If $\min(X_1,\ldots,X_p)$ and $\min(Y_1,\ldots,Y_q)$ have identical distributions then $p = q$ and there exists a permutation $(k_1,\ldots,k_p)$ of $(1,\ldots,p)$ such that $q_i = f_{k_i}$ $(i = 1,\ldots,p)$.

2.3 Dependent Random Variables

In the case of dependent competing risks Peterson (1976) has obtained bounds on the unobservable marginal survival probabilities $S_i(\cdot)$ in terms of observable crude survival probabilities $S_i^*(\cdot)$, in the case $p = 2$. He also obtains a bound on the joint survival function $\bar{F}(x_1,x_2) = P(X_1 > x_1,\ X_2 > x_2)$ in the following theorem.

THEOREM 4: (Peterson (1976))

Let $S_i(x) = P(X_i > x)$, $S_i^*(x) = P(X_i > x,\ \min(X_1,X_2) = X_i)$, $p_1 = P(X_1 < X_2)$, and $p_2 = P(X_2 < X_1)$ then

$$\bar{F}(x_1,x_2) = S_1^*(x_1) + S_2^*(x_2) - P(x_1 < X_1 < X_2 < x_2) \qquad \text{if } x_1 < x_2$$

$$= S_1^*(x_1) + S_2^*(x_2) - P(x_2 < X_2 < X_1 < x_1) \qquad \text{if } x_1 > x_2 \ .$$

Also

$$S_1^*[\max(x_1,x_2)] + S_2^*[\max(x_1,x_2)] \leq \bar{F}(x_1,x_2) \leq S_1^*(x_1) + S_2^*(x_2)$$

and

$$S_1^*(x_1) + S_2^*(x_1) \leq S_1(x_1) \leq S_1^*(x_1) + p_2 \quad ,$$

$$S_1^*(x_2) + S_2^*(x_2) \leq S_2(x_2) \leq S_2^*(x_2) + p_2 \quad .$$

Basu and Ghosh (1978) prove the following result which can be used in the identifiability problem.

THEOREM 5: (Basu and Ghosh (1978))

Let $\bar{F}_i(x_1,x_2) = \frac{\delta}{\delta x_i} \bar{F}(x_1,x_2)$, $(i=1,2)$ and let $f(x_1,x_2)$ be the joint p.d.f. of (X_1,X_2). Assume that $f(x_1,x_2) > 0$ for all (x_1,x_2) and

$$\int_{-\infty}^{\infty} - \bar{F}_i(z,z) \, (\bar{F}(z,z))^{-1} \, dz = \infty, \quad (i=1,2).$$

Define

$$\bar{G}_i(x) = \exp\{- \int_{-\infty}^{x} - \bar{F}_i(z,z) \, (\bar{F}(z,z))^{-1} \, dz\}$$

$$= 1 - G_i(x) \quad , \quad (i=1,2) \quad ,$$

then $G_i(\cdot)$ is a C.D.F. Let Y_i be independent and follow the distribution $G_i(\cdot)$, so that $\bar{G}(x_1,x_2) = \bar{G}_1(x_1)\,\bar{G}_2(x_2)$, then (U,I) has the same distribution whether (X_1,X_2) follow $F(x_1,x_2)$ or (X_1,X_2) follow $G(x_1,x_2)$.

In the case of dependence the identifiability problem makes sense for specified parametric distributions. Basu and Ghosh (1978 and 1980) have results for the bivariate normal distribution and multivariate exponential distributions (c.f. Block and Basu (1974), Marshall and Olkin (1967), Gumbel (1960)). Results for a general p follow readily for the exponential case.

3. Converting Dependent Models to Independent Ones

3.1 Basic Theorems

Implications of the result in Theorem 5 have also been considered by Miller (1977), Tsiatis (1975), Langberg, Proschan and Quinzi (1977,1981), and others. We state some of these results along with their implications.

As in the previous section let $\underline{X} = (X_1, \ldots, X_p)$ be a vector of positive random variables and let $U = \min(X_1, \ldots, X_p)$ be the observable system life. Let $\mathcal{I}$ denote the collection of nonempty subsets of $\{1, \ldots, P\}$. Let $\underline{H}$ be the vector of component life lengths in a series system of $(2^P - 1)$ components with system life T where the coordinates of H are indexed lexiographically by $I \in \mathcal{I}$. Define the failure patterns by

$$\xi(\underline{X}) = \begin{cases} I \text{ if } U = X_i \text{ for each } i \in I \text{ and } U \neq X_i \text{ for each } i \notin I \\ \emptyset \qquad \text{otherwise} \end{cases}$$

and

$$\xi^*(\underline{H}) = \begin{cases} I \text{ if } H_I < H_J \text{ for each } J \neq I \\ \emptyset \qquad \text{otherwise} \end{cases} .$$

We say $\underline{X}$ and $\underline{H}$ are equivalent in life length and failure pattern ($\underline{X} \overset{LP}{=} \underline{H}$) if $P(T > t,\ \xi^*(H) = I) = P(X > t,\ \xi(\underline{X}) = I),\ t \geq 0,\ I \in \mathcal{I}$. When $\underline{X} \overset{LP}{=} \underline{H}$ then the two system lifetimes are identically distributed and corresponding failure patterns have the same chance of occurring.

Langberg et al (1978) give necessary and sufficient conditions for replacing a set of dependent life lengths $\underline{X}$ by a set of independent life lengths $\underline{H}$ such that $\underline{X} \overset{LP}{=} \underline{H}$ in the following theorem.

THEOREM 6: (Langberg et al (1978))

Let $U = \min(X_1, \ldots, X_p)$ denote the life length of a p-component series system, where X_i represents the life length of the ith component, $i = 1, \ldots, p$. Define $\overline{F}_I(u) = P(U > u,\ \xi(\underline{X}) = I)$, $F_I(u) = P(U \leq u,\ \xi(\underline{X}) = I)$, $\overline{F}(u) = P(U > u)$ and $\alpha(F) = \sup\{u : \overline{F}(u) > 0\}$. Then the following statements hold:

(i) A necessary and sufficient condition for the existence of a set of independent random variables $(H_I,\ I \varepsilon \mathcal{I})$ which satisfy $\underline{H} \overset{LP}{=} \underline{X}$ is that the set of discontinuities of F_I be pairwise disjoint on the interval $[0,\alpha(F))$.

(ii) The distribution of $\{H_I,\ I \varepsilon \mathcal{I}\}$ in (i) are uniquely determined on the interval $[0,\alpha(F))$ as follows:

(1)
$$\overline{G}_I(t) = P(H_I > t) = \exp[-\int_0^t dF_I^c / \overline{F}] \times \prod_{a(I,j) \leq t} \{\overline{F}(a(I,j)) / [\overline{F}(a(I,j)) + f_I(a(I,j))]\} ,$$

$0 \leq t < \alpha(F)$, where F_I^c is the continuous part of F_I, $\{a(I,j)\}$ is the set of discontinuities of F_I and $f_I(a(I,j))$ is the size of the jump of F_I at $a(I,j)$.

Langberg et al (1981) show how the marginal survival functions of the dependent system can be recovered from the equivalent in LP independent system. Let $S_I(\cdot)$ denote the marginal survival function of $\underline{X}_I = \{X_i,\ i \varepsilon I\}$.

THEOREM 7: (Langberg et al (1981))

Let $X_1,\ldots,X_p$ be non-negative random variables such that the functions $F(\cdot,\ \xi(\underline{X}) = I)$ have no common discontinuities. Let $I \varepsilon \mathcal{I}$. Then for each $t \varepsilon [0,\alpha(F)]$, $\overline{S}_I(t) = \prod_{J \varepsilon I} \overline{G}_J(t)$ if and only if the following two conditions hold:

(C1)
$$\frac{S_I(a)}{S_I(a-)} = \begin{cases} \overline{F}(a)/\overline{F}(a-) & \text{for a discontinuity point of } F(\cdot,\xi(\underline{X}) = I) \\ 1 & \text{otherwise} \end{cases}$$

and

(C2) $P(X_{I'} \geq t | X_I = t) = P(X_{I'} > t | X_I > t)$, where I' is the complement of I in $\mathcal{I}$ and $\overline{G}$ is defined by (1).

EXAMPLE 2: Let $\underline{X} = (X_1, X_2)$ have the bivariate Weibull distribution described by Lee and Thompson (1974) with joint survival function

$$\bar{F}(x_1,x_2) = \exp(-\lambda_1 x_1^{\alpha_1} - \lambda_2 x_2^{\alpha_2} - \lambda_{12}[\max(x_1,x_2)]^{\alpha_{12}}) .$$

Applying Theorem 6 it follows that $\underline{X} \stackrel{LP}{=} (H_1, H_2, H_{12})$ where the H_i's are independent Weibull random variables with survival functions $\bar{G}_i(t) = \exp(-\lambda_i t^{\alpha_i})$, $i = 1,2,12$. The conditions of Theorem 7 are met so that $X_i = \min(H_i, H_{12})$, $i = 1,2$.

3.2 Constant Sum Models

Let X_1 and X_2 be positive random variables representing the failure time and censoring time of an individual under study. In the random censorship model we observe $U = \min(X_1,X_2)$ and $I = 1$ if $U = X_I$ (a death) or $I = 2$ if $U = X_2$ (a loss). Williams and Lagakos (1977) have examined conditions on the joint distribution of X_1 and X_2 under which the likelihood based on n observations on (U,I) is independent of the censoring distribution of X_2. Let $a(t) = P(I = 1 | t \leq X_1 < t + dt)$ and $dB(t) = P(t \leq U < t + dt, I = 2 | U \geq t)$. A model (X_1,X_2) is said to be a constant sum model if and only if

$$a(t) + \int_0^t dB(y) = 1 .$$

Kalbfleisch and McKay (1979) give an equivalent characterization of the constant sum condition in the following theorem.

THEOREM 8:

A model (X_1,X_2) is constant sum if and only if

$$P(t \leq X_2 < t + dt | U \geq t) = P(t \leq X_1 < t + dt | X_1 \geq t) .$$

We prove a sufficient condition for the constant sum model using the results of Langberg et al (1981).

THEROEM 9: A model (X_1,X_2) is constant sum if the set of discontinuities of $F_I(\cdot)$ are pairwise disjoint for all $I \in \mathcal{I}$, and ,

$$P(X_1 \geq x | X_2 = x) = P(X_1 > x | X_2 > x) \quad .$$

PROOF:

By Theorems 6 and 7, $(X_1 X_2) \overset{LP}{=} (H_1,H_2,H_{12})$ where the $H_{i's}$ are independent and $U = \min(H_1,H_2,H_{12})$ and $X_1 = \min(H_1,H_{12})$. Now

$$P(t \leq X_1 < t+dt | U \geq t) = P(t \leq X_1 < t+dt,\ U \geq t) \,/\, P(U \geq t)$$

$$= \frac{P(t \leq \min(H_1,H_{12}) < t+dt,\ \min(H_1,H_2,H_{12}) \geq t)}{P(\min(H_1,H_2,H_{12}) \geq t)}$$

$$= \frac{P(t \leq \min(H_1,H_{12}) < t+dt,\ H_1 \geq t,\ H_{12} \geq t)\ P(H_2 \geq t)}{P(H_1 \geq t,\ H_2 \geq t)\ P(H_2 \geq t)}$$

$$= P(t \leq X_1 < t+dt,\ X_1 \geq t) \,/\, P(X_1 \geq t)$$

$$= P(t \leq X_1 < t+dt | X_1 \geq t) \quad .$$

The result now follows by Theorem 8.

3.3 Inference When There is a Dependent Censoring Mechanism

The above result suggests using Theorem 6 to justify standard nonparametric techniques developed for censored data, under the assumption of an independent censoring mechanism, when the censoring mechanism is dependent but satisfies the conditions of Theorem 7. As an example consider the two sample problem. Let (X_1,X_2) and (Y_1,Y_2) be bivariate positive random variables. Suppose the marginal survival function of X_1 is $S_1(t)$ and the marginal survival function of Y_1 is $R_1(t)$ and X_2 and Y_2 are possibly dependent censoring variables. Observations on (X_1,X_2) consist of observing $\min(X_1,X_2)$ and the failure pattern $\xi(X_1,X_2)$. Similarly, for (Y_1,Y_2), we observe $\min(Y_1,Y_2)$ and $\xi(Y_1,Y_2)$. The problem is to test $H_0{:}S_1(t) = R_1(t)$, $t \geq 0$ based on n_1 observations on (X_1,X_2) and n_2 observations on (Y_1,Y_2). Suppose the conditions of Theorem 7 hold for both

$\underline{X}$ and $\underline{Y}$. Then $(X_1,X_2) \overset{LP}{=} (H_1,H_2,H_{12})$ and $(Y_1,Y_2) \overset{LP}{=} (K_1,K_2,K_{12})$ where the H_i's and K_i's are independent with survival functions $\overline{G}_i$ and $\overline{M}_i$ $(i=1,2,12)$, respectively. Also $S_1(t) = \overline{G}_1(t)\ \overline{G}_2(t)$ and $R_1(t) = \overline{M}_1(t)\ \overline{M}_{12}(t)$. Hence, testing $S_1(t) = R_1(t)$ is equivalent to testing $H_0: \overline{M}_1(t)\ \overline{M}_{12}(t) = \overline{G}_1(t)\ \overline{G}_{12}(t)$. Observations with $\xi(X_1,X_2) = \{1\}$ or $\{1,2\}$ give complete information about $S_1(t)$. Similarly for $R_1(t)$. Those with $\xi(X_1,X_2) = \{2\}$ are censored for testing H_0: $S_1(t) = R_1(t)$ but now from independent censoring distributions. Thus standard nonparametric techniques for independent censoring variables such as Gehan (1965) or Efron (1966) may be used to test this hypothesis.

3.4 Estimating Joint Survival

Theorem 7 can be used to obtain a consistent estimator of the joint survival function, $\overline{F}(x_1,x_2)$, of a bivariate random variable (X_1,X_2). Suppose the conditions of Theorem 7 hold, then $(X_1,X_2) \overset{LP}{=} (H_1,H_2,H_{12})$ and $X_1 = \min(H_1,H_{12})$, $X_2 = \min(H_2,H_{12})$ with H_1,H_2,H_{12} independent. Now

$$\begin{aligned}
\overline{F}(x_1,x_2) &= P(X_1 > x_1,\ X_2 > x_2) \\
&= P(\min(H_1,H_{12}) > x_1,\ \min(H_2,H_{12}) > x_2) \\
&= P(H_1 > x_1,\ H_2 > x_2,\ H_{12} > \max(x_1,x_2)) \\
&= P(H_1 > x_1)\ P(H_2 > x_2)\ P(H_{12} > \max(x_1,x_2) \\
&= \overline{G}_1(x_1)\ \overline{G}_2(x_2)\ \overline{G}_{12}(\max(x_1,x_2)) \quad .
\end{aligned}$$

Let $\hat{\overline{G}}_i(t)$ be a consistent estimator of $\overline{G}_i(t)$ for $i = 1,2,12$. A consistent estimator of $\overline{F}(x_1,x_2)$ is given by

$$\hat{\overline{F}}(x_1,x_2) = \hat{\overline{G}}_1(x_1)\ \hat{\overline{G}}_2(x_2)\ \hat{\overline{G}}_{12}(\max(x_1,x_2)) \quad .$$

4. Accelerated Life Testing and Safe Dose Estimation Under Competing Causes of Failure

Accelerated life testing of a product under more severe than normal conditions is commonly used to reduce test time and costs. Data collected at such accelerated conditions are used to obtain estimates of the parameters of a stress translation function. This function is then used to make inference about product life under normal operating conditions.

Klein and Basu (1981a,b,c) have considered the problem of accelerated life tests when the product of interest is a p component series system. Each of the components is assumed to have either exponential distributions or Weibull distributions with different or the same shape parameter.

Klein and Basu considered the following model. Let $X_1,\dots,X_p$ denote the component life lengths in a p component series system. At a constant application of a stress $V_i (i = 1,\dots,s)$ assume that the jth component has hazard rate given by

$$h_j(x,v_i;\ \underline{\alpha}_j,\ \underline{\beta}_j) = g_j(x,\ \underline{\alpha}_j)\ \lambda_j(V_i,\ \underline{\beta}_j),\quad i = 1,\dots,s$$
$$j = 1,\dots,P$$

as introduced in Klein and Basu (1981d). The α_j's may vary from component to component to allow for different component reliabilities. For $\lambda_j(V,\ \underline{\beta}_j)$ assume a model of the form

$$\lambda_j(V,\ \underline{\beta}_j) = \exp\left(\sum_{\ell=0}^{k_j} \beta_{j\ell}\ \Theta_{j\ell}(V)\right) \tag{2}$$

where $\Theta_{j0}(V) = 1$ and $\Theta_{j1}(V),\dots,\Theta_{jk_j}(V)$ are non-decreasing functions of V. The $\Theta_j(\cdot)$'s may differ from one component to another.

This model includes the standard models, namely, the power rule with $\lambda_j(V,\ \underline{\beta}_j) = \beta_{j0}V^{\beta_{j1}}$, the Arrhenius reaction rate model with $\lambda_j(V,\ \underline{\beta}_j) = \exp(\beta_{j0} - \beta_{j1}/V)$, and the Eyring model for a single stress with

$\lambda_j(V, \underline{\beta}_j) = V^{\beta_{j1}} \exp(\beta_{j0} - \beta_{j2}/V)$, as special cases.

The model can be derived from the interpretation of the effects of a carcinogen on a cell as proposed by Armitage and Doll (1961). For details see Klein and Basu (1981d). To produce cancer in a single cell k independent events must occur. The effects of an increased dose of a carcinogen is to increase the rate at which these k events occur. If, for the jth disease, this increase is of the form $\exp(\beta_{j\ell}\Theta_{j\ell}(V))$, $(\ell = 1, \ldots, k_j)$ the model (2) is obtained. If this increase is linear in V the model of Hartley and Sielkin (1977) is obtained. Thus their model is a first order Taylor series approximation to (2). Klein and Basu (1981c) have extended the results of Hartley and Sielkin (1977) to the competing risks model.

ACKNOWLEDGEMENT

This research was supported in part by the Office of Naval Research, Grant No. N00014-78-C-0655.

REFERENCES

Anderson, T.W. and Ghurye, S.G. (1977). Identification of parameters by the distribution of a maximum random variable. Journal of the Royal Statistical Society B 39, 337-342.

Armitage, P. and Doll, R. (1961). Stochastic models for carcinogens. Proceedings of the Fourth Berkeley Symposium in Mathematical Statistics, Vol. IV University of California Press, 19-38, Berkeley, California.

Basu, A.P. (1981a). Identifiability problems in the theory of competing and complementary risks - a survey. Statistical Distributions in Scientific Work (Taillie, Patil, and Baldesaari, Eds.). Dorerecht, Holland: Reidel Publishing Company, 335-348.

Basu, A.P. (1981b). Identifiability. Encyclopedia of Statistics (Johnson and Kotz, Eds.). To appear.

Basu, A.P. and Ghosh, J.K. (1978). Identifiability of multinormal and other distributions under competing risks models. Journal of Multivariate Analysis **8**, 417-429.

Basu, A.P. and Ghosh, J.K. (1980). Identifiability of distributions under competing risks and complementary risks models. Communications in Statistics, Theory and Methods A **9**(14), 1515-1525.

Basu, A.P. and Ghosh, J.K. (1981). Generalized competing risks theory. Unpublished.

Berman, S.M. (1963). Note on extreme values, competing risks and semi_Markov processes. Annals of Mathematical Statistics **34**, 1104-1106.

Block, H.W. and Basu, A.P. (1974). A continuous bivariate exponential extension. Journal of the American Statistical Association **69**, 1031-1037.

Efron, B. (1967). The two sample problem with censored data. Proceedings of the Fifth Berkeley Symposium in Mathematical Statistics, Vol. IV, University of California Press, 831-853, Berkeley, California.

Gehan, E.A. (1965). A generalized Wilcoxon test for comparing arbitrarily singly-censored samples. Biometrika **52**, 203-223.

Gumbel, E.J. (1966). Bivariate exponential distributions. Journal of the American Statistical Association **55**, 698-707.

Hartley, H.O. and Sielkin, R.L., Jr. (1977). Estimation of "safe doses" in carcinogenesis experiments. Biometrika **33**, 1-30.

Kalbfleisch, J.D. and MacKay, R.J. (1979). On constant-sum models for censored survival data. Biometrika **66**, 87-96.

Kaplan, E.L. and Meier, P. (1958). Nonparametric estimation from incomplete observations. Journal of the American Statistical Association **53**, 457-481.

Klein, J.P. and Basu, A.P. (1981a). Weibull accelerated life tests when there are competing causes of failure. Communications in Statistics, Theory and Methods A **10**(20),2073-2100.

Klein,J.P. and Basu, A.P. (1981b). Accelerated life testing under competing exponential failure distributions. IAPQR Transactions. To appear.

Klein, J.P. and Basu, A.P. (1981c). Accelerated life testing under competing Weibull causes of failure (submitted for publication).

Klein, J.P. and Basu, A.P. (1981d). A model for accelerated life testing and low dose estimation (submitted for publication).

Klein, J.P. and Basu, A.P. (1981e). Low dose estimation when there are multiple diseases. Technical Report No. 247, Department of Statistics, The Ohio State University.

Langberg, N., Proschan, F. and Quinzi, A.J. (1978). Converting dependent models into independent ones, preserving the essential features. Annals of Probability **6**, 174-181.

Langberg, N., Proschan, F. and Quinzi, A.J. (1981). Estimating dependent life lengths, with applications to the theory of competing risks. Annals of Statistics **9**, 157-167.

Lee, L. and Thompson, W.A. (1974). Reliability of multiple component systems. Transactions of the Seventh Prague Conference on Information Theory, Statistical Decision Functions, Random Processes and of the 1974 European Meeting of Statisticians, 329-336.

Miller, D.R. (1977). A note on independence of multivariate life times in competing risk models. Annals of Statistics **5**, 576-579.

Marshall, A.W. and Olkin, I. (1967). A multivariate exponential distribution. Journal of the American Statistical Association **62**, 30-40.

Peterson, A.V. (1976). Bounds for a joint distribution function with fixed sub-distribution functions: application to competing risks. Proceedings of the National Academy of Sciences **73**, 11-13.

Peterson, A.V. (1977). Expressing the Kaplan-Meier estimator as a function of empirical subsurvival functions. Journal of the American Statistical Association **72**, 854-858.

Tsiatis, A. (1975). A nonidentifiability aspect of the problem of competing risks. Proceedings of the National Academy of Sciences **72**, 20-22.

Williams, J.S. and Lagakos, S.W. (1977). Models for censored survival analysis; constant sum and variable sum models. Biometrika **64**, 215-224.

FREUND'S BIVARIATE EXPONENTIAL DISTRIBUTION AND CENSORING

Sue Leurgans and Wei-Yann Tsai

University of Wisconsin-Madison

and

John Crowley

University of Washington and Fred Hutchinson Cancer Research Ctr.

0. SUMMARY

In some problems, a bivariate random vector (T_1,T_2) with bivariate cumulative distribution function F is observed for each of n independent subjects, but the coordinates may be subject to censoring. In the first section, we describe several mechanisms which can generate the censorship. The usual nonparametric approaches to estimation of F are then shown to be unsatisfactory. Therefore, in the third section, we describe a parametric model due to Freund (1961). This model is studied not because all data can be forced to fit this specific parametric form, but because this model suggests some approaches to the nonparametric problem. These ideas, together with some relationships to the work of other authors, are outlined in the fourth section.

1. Bivariate Times and Censoring Patterns

In this section, we outline several mechanisms which can generate bivariate censored times. A distinction between univariate and bivariate censoring is developed.

Univariate censoring arises naturally in two similar contexts. Firstly, the experimental units may contain two similar components (such as ears, elbows, knees, kidneys or engines) whose survival is being studied. Alternatively, the experimental units may contain two dissimilar components whose survival is being studied. Unlike competing risk problems, neither component is essential for the survival of the experimental unit. In both cases, censoring occurs when the experimental unit is removed from observation before both components have been observed to fail. Examples include lifetimes of pumps and hoses on 15 tractors given in Barlow and Proschan (1977) and the times of first responses to treatment (as observed at one site, perhaps a head or a tumor) and times of first sign of toxicity or death, as discussed by Lagakos (1976).

In the examples above, all times for any experimental unit are measured on a single clock from a common origin. However, double clocks are natural when studying the times required from initiation of treatment until the first sign of response in two successive courses of treatment in the same patient (see Gross and Lam (1981)) or the lifetimes of two paired subjects (siblings or other kin). The time until response to treatment and the length of the subsequent disease-free interval also requires two separate clocks. Indeed, the random variables need not be times in the usual sense. Variables could be cumulative dose or cumulative cost. Censoring would occur when an experimental unit (or component) is removed from observation for reasons independent of both responses.

To model the censoring, independent censoring vectors (C_1,C_2) are postulated to exist for each bivariate vector (T_1,T_2). We suppose that the vectors (C_1,C_2) form a sample from a bivariate distribution G. While such an assumption will not always be valid, it permits censoring times to differ. The observed quantities are then $X_i = \min(T_i,C_i)$ and $D_i = [T_i \leq C_i]$ $(i=1,2)$. (The symbol $[A]$ denotes the indicator function of the event A.)

When a single clock governs both times, censoring occurs when an experimental unit is removed from observation. Since C_1 will always equal C_2 in this case, this censoring structure will be referred to as univariate censoring. When the censoring times can differ, the censoring will be called bivariate.

The distinction between univariate and bivariate censoring is clear when the observations (X_1,X_2) are plotted in R^2. If $D_i = 0$, the ith coordinate is censored and an arrow parallel to the ith axis is drawn from (X_1,X_2). If $D_i = 1$, the ith coordinate is observed exactly, and the arrow is omitted. If the censoring is bivariate, the diagram can resemble Figure 1.

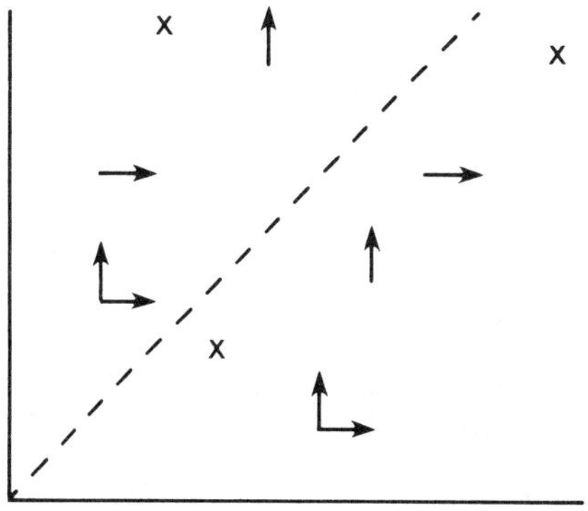

Figure 1: Schematic Diagram for Bivariate Censoring

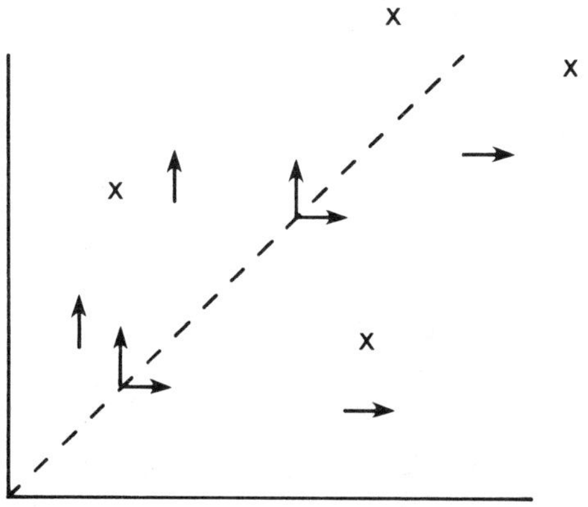

Figure 2: Schematic Diagram for Univariate Censoring

However, if the censoring is univariate, Figure 1 is impossible. In univariate censoring, if exactly one coordinate is censored, it must be the coordinate with the larger value. Consequently, if a point has one arrow attached, that arrow must point away from the diagonal lime $X_1 = X_2$. Furthermore, if both coordinates are censored, since the censoring variable must be the same for each component, the two coordinates must have the same value and all points with two arrows must be based on the diagonal. Figure 2 indicates a possible diagram when univariate censoring is present. Both of these situations are clearly different from competing risks problems, when at most one lifetime can be observed exactly on each experimental unit and observation of the surviving component is censored at the end of the first lifetime, resulting in a diagram similar to Figure 3:

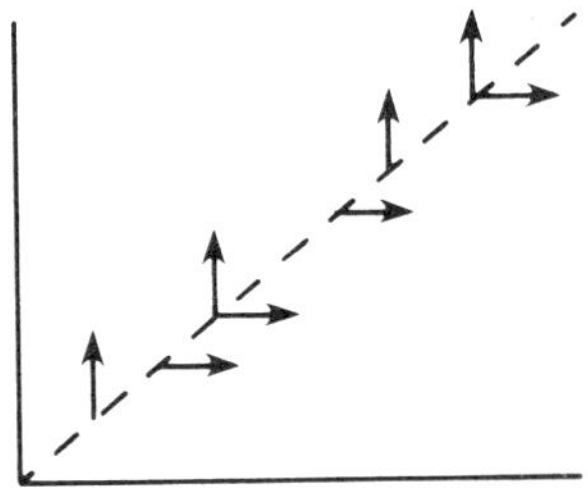

Figure 3: Schematic Diagram for Competing Risks with (Univariate) Censoring

2. Nonparametric Approaches

The principles of generalized maximum likelihood estimation and of self-consistency are often used to derive the product limit estimator of the cumulative distribution function of a single survival time in the presence of random censoring. In this section, these principles are shown to be inadequate for estimation of bivariate cumulative distribution functions.

In regular parametric settings, maximum likelihood estimators are well-known to have optimal asymptotic properties. The likelihood can be viewed as the Radon-Nikodym derivative of a parametrized probability measure with respect to a carrier measure. Since Radon-Nikodym derivatives can often be computed even when the "parameter" is not finite dimensional and a likelihood is not defined, Kiefer and Wolfowitz (1956) suggested a Generalized Maximum Likelihood Estimator (GMLE) for nonparametric problems. In parametric settings, the GMLE reduces to the usual maximum likelihood estimator. However, the generalized maximum likelihood principle is not known to guarantee any optimal properties, as occurs in the finite dimensional case. Johansen (1978) showed that the product limit estimator of Kaplan-Meier (1957) is the GMLE of F in the class of all CDF's.

Another property of the product limit estimator was established by Efron (1967), who named the property self-consistency. In the univariate problem, an estimator $\hat{F}$ is said to be self-consistent if

$$1 - \hat{F}(t) = \sum_{i=1}^{n} \frac{d_i[t_i \geq t]}{n} + \sum_{i=1}^{n} \frac{(1-d_i)}{n} \frac{1-\hat{F}(t)}{1-\hat{F}(t_i)} ,$$

or, the proportion estimated to survive past t is equal to the proportion of the subjects observed to survive past t plus the sum for all individuals censored before t of the estimated conditional probability of surviving past t, given survival to the censoring time. Efron showed that, up to possible indeterminacy for $t \geq t_{(n)}$, the only self-consistent estimator of the cumulative distribution function is the product limit estimator. Thus the GMLE is self-consistent.

In a 1980 Stanford Ph.D. dissertation, Muñoz studied nonparametric estimation of a bivariate distribution function in the presence of univariate censoring. He showed that the GMLE is self-consistent. He also showed that the GMLE is supported on three kinds of sets: points, rays and regions. The points of support are those (X_{1i},X_{2i}) with $(D_{1i},D_{2i}) = (1,1)$. The rays of support are sets $\{(x,y): x = X_{1i},\ y \geq X_{2i}\}$ with $(D_{1i},D_{2i}) = (1,0)$ or $\{(x,y): x \geq X_{1i},\ y = X_{2i}\}$ with $(D_{1i},D_{2i}) = (0,1)$. One region of support may exist: $\{(x,y): x \geq X_{1i},\ y \geq X_{2i}\}$ will be a region of support if $(D_{1i},D_{2i}) = (0,0)$ and the region contains no other points, rays or regions. Thus the support of the GMLE is the minimal set in which the true times corresponding to observed times must lie.

Muñoz showed that the mass of each set is determined, but the distribution of the mass within the set is completely arbitrary. Since, under random censorship, a non-negligible proportion of the observations will be censored in a single component, a non-negligible proportion of the mass is not located by the GMLE. Therefore there are self-consistent estimators of bivariate distribution functions which do not converge to the correct limit. The fact that self-consistency alone is inadequate is recognized in the calculations of Muñoz's example, despite a theorem which states that bivariate self-consistent estimators are asymptotically consistent. However, Campbell (1981) establishes that self-consistent estimators of discrete distributions are asymptotically consistent.

3. Freund's Model

Since the completely nonparametric approaches outlined above are unsatisfactory, a simple parametric model introduced by Freund (1961) will be described in this section. We will show that the resulting joint density for (T_1,T_2) is a tractable curved exponential family when univariate censoring is present. Subfamilies described by Block and Basu (1974) and by Lagakos (1976) are seen to be much less tractable under censoring.

3.1 Freund's Distribution

We suppose that the pair of times being studied can be recorded from a single clock. The experimental unit can be thought of as being under continuous observation, changing state whenever clocktime passes T_1 or T_2. If T_1 and T_2 are jointly absolutely continuous, the states and transitions possible at time t are indicated in Figure 4.

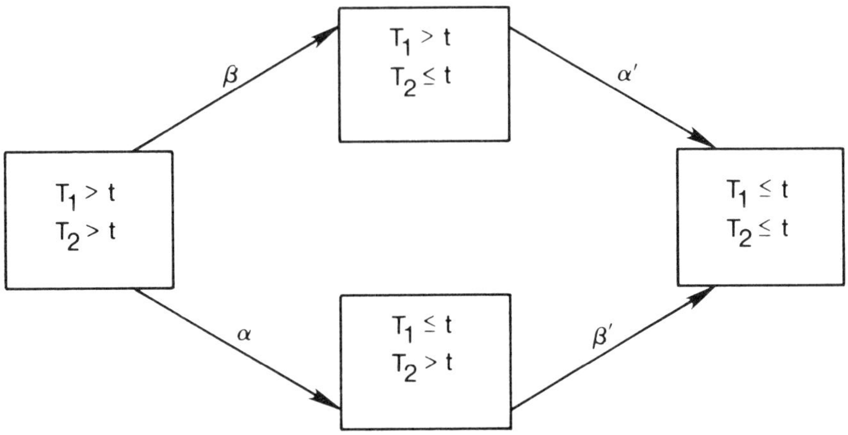

Figure 4: States and Transitions for Freund Model

If the Markov property is assumed, the times between transitions will have exponential distributions with the four positive parameters indicated in the diagram. Since the Markov property implies that the difference between the two times must be independent of the exact value of the smaller random variable, this model gives the following joint density for T_1 and T_2:

$$(1) \qquad f_{T_1,T_2}(t_1,t_2) = \begin{cases} \alpha e^{-(\alpha+\beta)t_1} \beta' e^{-\beta'(t_2-t_1)} & 0 < t_1 < t_2 \\ \beta e^{-(\alpha+\beta)t_2} \alpha' e^{-\alpha'(t_1-t_2)} & 0 < t_2 < t_1 \end{cases} .$$

This density was introduced by Freund (1961), who showed that the marginal distributions are not exponential. Freund also calculated the expectations, variances and covariance of T_1 and T_2. He showed that the correlation coefficient need not be non-negative, but can range from -1/3 to 1.

3.2 Inference

In this sub-section, we show that Freund's distribution is a curved exponential family under univariate censoring and derive the closed form maximum-likelihood estimators. Bivariate censoring causes the dimension of the statistic to be random (and stochastically increasing with n). At the end of this sub-section a simpler alternative to the maximum-likelihood estimator is suggested for bivariate censoring.

In the presence of univariate censoring with density g and distribution function G, the likelihood for Freund's model is

$$(2) \qquad e^{\eta^T(\alpha,\beta;\alpha',\beta')\, z_+(\underset{\sim}{x},\underset{\sim}{d})} \prod_{i=1}^{n} \left[(1-G(x_i)) \left(\frac{g(\underset{\sim}{x}_i)}{1-G(\underset{\sim}{x}_i)} \right)^{1-d_{1i}d_{2i}} \right]$$

where

$$\eta(\alpha,\beta;\alpha',\beta') = \begin{pmatrix} -(\alpha+\beta) \\ -\alpha' \\ -\beta' \\ \ell n(\frac{\alpha\beta'}{\alpha'\beta}) \\ \ell n(\alpha'/\alpha) \\ \ell n\ \alpha \\ \ell n\ \beta \end{pmatrix}, \quad z(\underset{\sim}{x},\underset{\sim}{d}) = \begin{pmatrix} \min(x_1,x_2) \\ (x_1-x_2)_+ \\ (x_2-x_1)_+ \\ [x_1 < x_2]d_1d_2 \\ d_1d_2 \\ d_1 \\ d_2 \end{pmatrix}$$

$z_+(\underset{\sim}{x},\underset{\sim}{d}) = \sum_{i=1}^{n} z(\underset{\sim}{x}_i,\underset{\sim}{d}_i)$, and $(x)_+ = x[x>0]$. Because four of the seven coordinates of η are non-linear functions of the parameters, this is a four-parameter curved exponential family with a seven-dimensional sufficient statistic (see Efron (1978)). If no censoring is present, $G \equiv 0$ and the last three components of z_+ are all equal to n. Therefore $\eta^T(\alpha,\beta;\alpha',\beta')\ z_+(\underset{\sim}{x},\underset{\sim}{d})$ is an affine function of four sufficient statistics, and Freund's distribution is a regular exponential family in the absence of censoring. In either case, the theory of exponential families implies that the maximum-likelihood estimator is given by the solution of

$$\begin{pmatrix} -1 & 0 & 0 & 1/\alpha & -1/\alpha & 1/\alpha & 0 \\ -1 & 0 & 0 & -1/\beta & 0 & 0 & 1/\beta \\ 0 & -1 & 0 & -1/\alpha' & 1/\alpha' & 0 & 0 \\ 0 & 0 & -1 & 1/\beta' & 0 & 0 & 0 \end{pmatrix} \frac{\underset{\sim}{z}_+}{n} = \underset{\sim}{0} \quad .$$

If all components z are positive, the maximum-likelihood estimator is obtained by solving four one-parameter equations. Each of the resulting estimators is the ratio of a number of occurrences to a total exposure time:

$$\hat{\alpha} = \frac{\sum_{i=1}^{n} D_{1i}(1-D_{2i}[T_{1i} \geq T_{2i}])}{\sum_{i=1}^{n} \min(T_{1i},T_{2i})}$$

$$\hat{\beta} = \frac{\sum_{i=1}^{n} D_{2i}(1-D_{1i}[T_{1i} < T_{2i}])}{\sum_{i=1}^{n} \min(T_{1i},T_{2i})}$$

(3)

$$\hat{\alpha}' = \frac{\sum_{i=1}^{n} D_{1i}D_{2i}[T_{1i} > T_{2i}]}{\sum_{i=1}^{n} (T_{1i} - T_{2i})_{+}}$$

$$\hat{\beta}' = \frac{\sum_{i=1}^{n} D_{1i}D_{2i}[T_{1i} < T_{2i}]}{\sum_{i=1}^{n} (T_{2i} - T_{1i})_{+}} .$$

In the absence of censoring, these estimators reduce to those obtained by Freund (1961). It is clear that these estimators always take values inside the parameter space. Furthermore, the exponential family form implies that UMP tests are possible for all one-sided alternatives that can be specified in terms of a single linear transformation of the natural parameters. Thus the best test based on complete observations of the null hypothesis $\alpha' = \beta' = \alpha+\beta$ (stress-passing) against $\alpha' = \beta' > \alpha+\beta$ (increased stress) will not depend on $\alpha+\beta$. Clearly, fewer UMP tests exist in the presence of univariate censoring. The strong consistency and joint asymptotic normality of the estimators follow routinely from the strong law of large numbers and the central limit theorem applied to iid vectors Z.

If bivariate censoring is present, then the log-likelihood is not always linear in the parameters. To see this, note that the likelihood factor for terms with $D = (1,0)$ and $X_1 > X_2$ is

$$\int_{X_2} f_{X_1X_2}(x_1,x_2)dx_2 = \frac{\beta\alpha'}{\alpha+\beta-\alpha'} e^{-\alpha'(X_1-X_2)} e^{-(\alpha+\beta)X_2} + e^{-(\alpha+\beta)X_1} \frac{(\alpha-\alpha')(\alpha+\beta)}{\alpha+\beta-\alpha'} .$$

This factor is not of exponential family form, but is a mixture of two exponential family densities, reflecting the fact that it may be unclear which parameters were acting on which experimental units. The product of terms of this type does not generate a sufficient statistic of fixed dimension. Since the sufficient statistic is more complex than that for univariate censoring or complete data, the solution of the maximum likelihood equations will generally be more difficult.

One way to simplify the estimation procedure can be thought of as modifying the observations to reflect the observations that would have been made if the censoring had been univariate. Formally, define

$$(X_1^*, X_2^*, D_1^*, D_2^*)^T = \begin{cases} \begin{pmatrix} \min(X_1,X_2) \\ \min(X_1,X_2) \\ 0 \\ 0 \end{pmatrix} & \text{if } \begin{cases} X_1 > X_2 \text{ and } D_1 = 0 \\ \text{or} \\ X_1 < X_2 \text{ and } D_2 = 0 \end{cases} \\ (X_1, X_2; D_1, D_2)^T & \text{otherwise} \end{cases} .$$

The estimators $\tilde{\alpha}, \tilde{\beta}, \tilde{\alpha}'$ and $\tilde{\beta}'$ are obtained by applying the estimators (3) to the univariately censored $\{(X_{1i}^*, X_{2i}^*; D_{1i}^*, D_{2i}^*),\ 1 \leq i \leq n\}$. While it is clear that this approach throws away information and cannot always be efficient, this approach does provide closed form consistent estimators which are approximately normal and independent in large samples. The precise efficiency properties remain to be determined.

3.3 Subfamilies

Several sub-models of Freund's distribution have been proposed. Block and Basu (1974) point out that a three-parameter subfamily of Freund's distribution corresponds to the absolutely continuous component of the bivariate exponential distribution derived by Marshall and Olkin (1967). The three parameters are a linear function of the first three coordinates of η in (2) and correspond to the constraint that $\beta(\beta' - \beta) = \alpha(\alpha' - \alpha)$. The non-linearity of this function and the resulting curvature of the exponential family are reflected in the fact that the maximum-likelihood estimators for λ_1, λ_2 and λ_{12} do not have closed-form expressions, even for complete data. The sub-family characterization does imply that the maximum-likelihood equations based on complete data (univariate censoring) are determined by a four (seven) dimensional sufficient statistic. (See Section 7 of Block and Basu (1974) for complete data equations.)

Lagakos (1976) presents a three-parameter family for joint analysis of response time and survival time in cancer treatment studies. With a convention that response times are not observed after death, this family corresponds to Freund's family with the restriction that $\beta=\beta'$, a non-linear constraint on the natural parameters.

Since neither family exhibits a compelling superiority over the Freund family, we suggest considering the full family whenever either subfamily is fitted.

4. Extensions

One way to extend Freund's model to a nonparametric family is to allow the parameters to be functions of time, permitting α' and β' to depend on the first failure times. This yields the functions $\alpha(t)$, $\beta(t)$, $\alpha'(t|y)$ and $\beta'(t|y)$, corresponding to $\lambda_1(t)$, $\lambda_2(t)$, $\lambda_{1|2}(t|y)$ and $\lambda_{2|1}(t|x)$ of Cox (1972). None of these functions correspond to hazard gradients. In his dissertation research, Mr. Tsai is investigating nonparametric estimators of these functions in the presence of censoring. When F is absolutely continuous, in order to obtain consistent estimators of $\alpha'(t|y)$ and $\beta'(t|x)$, some form of smoothing is required, since otherwise no more than one datum could be used to estimate each conditional function.

Other researchers have imposed additional structure. If the experimental units are assumed to have a non-stationary Markov property, $\alpha'(t|y) = \alpha'(t)$ and

$\beta'(t|x) = \beta'(t)$. Nonparametric tests for this model are described in Aalen, Borgan, Keiding and Thormann (1980) and in Voelkel (1980). In some cases, it is more reasonable to believe that when one component fails, the other component begins to age differently. The semi-Markov property requires instead that $\alpha'(t|y) = \alpha'(t-y)$ and $\beta'(t|x) = \beta'(t-x)$. Lagakos, Sommer and Zelen (1978) and Voelkel (1980) studied this model.

Freund's distribution has been extended to more than one time and to allow a positive probability that $T_1 = T_2$. For some such extensions and additional references, see Block (1975) and Proschan and Sullo (1975).

ACKNOWLEDGEMENT

This research was partially supported by NIH grant 2R01-CA-18332.

REFERENCES

Aalen, O., Borgan, O., Keiding, N., and Thormann, J. (1980). Interaction between life history events. Nonparametric analysis for prospective and retrospective data in the presence of censoring. Scandinavian Journal of Statistics 7, 161-171.

Barlow, R.E. and Proschan, F. (1977). Techniques for analyzing multivariate failure data, in Tsockos, C.P. and Shimi, I.N. eds., Theory and Applications of Reliability 1, 373-396.

Block, H.W. (1975). Continuous multivariate exponential extensions, in Barlow, Fussell and Singpurwalla, eds., Reliability and Fault Tree Analysis, SIAM: 285-306.

Block, H.W. and Basu, A.P. (1974). A continuous bivariate exponential extension. Journal of the American Statistical Association 69, 1031-1037.

Campbell, G. (1981). Nonparametric bivariate estimation with randomly censored data. Biometrika 68, 417-422.

Cox, D.R. (1972). Regression models and life-tables (with discussion). Journal of the Royal Statistical Society B 34, 187-220.

Efron, B. (1967). The two-sample problem with censored data. Proceedings of the Fifth Berkeley Symposium on Mathematical Statistics and Probability, Vol. IV, University of California Press, Berkeley, California, 831-853.

Efron, B. (1978). The geometry of exponential families. Annals of Statistics 6, 363-376.

Freund, J.E. (1961). A bivariate extension of the exponential distribution. Journal of the American Statistical Association 56, 971-977.

Gross, A.J. and Lam, C.F. (1981). Paired observations from a survival distribution. Biometrics 37, 505-512.

Johansen, S. (1978). The product limit estimator as maximum likelihood estimator. Scandinavian Journal of Statistics 5, 195-199.

Kaplan, E.L. and Meier, P. (1958). Nonparametric estimation from incomplete observations. Journal of the American Statistical Association 53, 457-481.

Kiefer, J. and Wolfowitz, J. (1956). Consistency of the maximum likelihood estimator in the presence of infinitely many incidental parameters. Annals of Mathematical Statistics 27, 887-906.

Lagakos, S.W. (1976). A stochastic model for censored-survival data in the presence of an auxiliary variable. Biometrics 32, 551-559.

Lagakos, S.W., Sommer, C.J. and Zelen, M. (1978). Semi-Markov models for partially censored data. Biometrika 65, 311-317.

Muñoz, A. (1980a). Nonparametric observation from censored bivariate observations. Technical Report, Division of Biostatistics, Stanford University.

Muñoz, A. (1980b). Consistency of the self-consistent estimator of the distribution function from censored observations. Technical Report, Division of Biostatistics, Stanford University.

Proschan, F. and Sullo, R. (1975). Estimating the parameters of a bivariate exponential distribution in several sampling situations, in Proschan, F. and Serfling, R.J., eds. Reliability and Biometry, SIAM, 1974, 423-440.

Voelkel, J. (1980). Multivariate counting processes and the probability of being in response function. Ph.D. dissertation, Statistics Department, University of Wisconsin-Madison.

ASYMPTOTIC PROPERTIES OF SEVERAL NONPARAMETRIC MULTIVARIATE DISTRIBUTION FUNCTION ESTIMATORS UNDER RANDOM CENSORING

Gregory Campbell

Laboratory of Statistical & Mathematical Methodology
National Institutes of Health

1. Introduction

The problem of nonparametric estimation of a multivariate distribution function in the presence of random censoring is considered. The multivariate lifetimes could represent the times to death of animals in fixed-sized litters, the failure times of components in a multicomponent system, the observations of participants of a matched triples study, or the onset times to stages of a disease in a patient. In the special bivariate case, there are the numerous examples of paired data on eyes, lungs, kidneys, twins or married couples. It is possible that the censoring is univariate or multivariate. Whereas the censoring of times to death of animals in litters born at random times yet truncated at a fixed time is an example of univariate censoring, the truncation at a fixed time of measures on the participants in a matched triple study would provide trivariate independent censoring. The study of lifelengths of twins and married couples would provide an example of bivariate censoring with possible dependence between the two censoring variables.

The estimation of one-dimensional distribution function estimators with randomly censored data has been extensively developed. The product-limit estimator was proposed by Kaplan and Meier (1958). Under suitable conditions, asymptotic normality and weak convergence of this estimator was established by Breslow and Crowley (1974) and strong uniform almost sure convergence was proved by Földes and Rejtö (1981).

The bivariate problem has merited some attention recently. Campbell (1981) estimated the bivariate distribution function under bivariate censoring for discrete or grouped data via the EM algorithm. Leurgans, Tsai, and Crowley (this volume) have proposed an estimator for univariate censoring that utilizes Freund's bivariate exponential distribution. Campbell and Földes (1982) have proposed several estimators based on hazard gradient estimators and on products of one-dimensional product-limit estimators. It is the weak convergence of the latter estimators which is the purpose of this paper.

A path-dependent distribution function estimator based on the hazard gradient is introduced in §2 after some notational development. The result of strong uniform almost sure convergence of the estimator, which was proved in Campbell and Földes (1982), is presented.

A topological discussion in §3 precedes an important lemma on empirical processes in two-dimensional time. The main theorems of §4 prove the weak convergence of the suitably normalized estimator. The discussion in the final section considers estimators with different paths as well as estimators which are products of product-limit estimators. The extension from two to k dimensions is noted.

2. Notation and the Estimator

For simplicity of exposition, bivariate observations are considered. Let $\{\underset{\sim}{X}_i\}_{i=1}^{\infty}$ denote a sequence of independent random variables, $\underset{\sim}{X}_i = (X_{i1}, X_{i2})$, from the continuous bivariate distribution function F. Each $\underset{\sim}{X}_i$ represents the lifetimes of a pair of (possibly dependent) items. Let $\{\underset{\sim}{C}_i\}_{i=1}^{\infty}$ denote a sequence of independent random variables, $\underset{\sim}{C}_i = (C_{i1}, C_{i2})$, from the continuous bivariate distribution function G. It is assumed that $\{\underset{\sim}{X}_i\}_{i=1}^{\infty}$ and $\{\underset{\sim}{C}_i\}_{i=1}^{\infty}$ are mutually independent.

In general $\underset{\sim}{X}_i$ and $\underset{\sim}{C}_i$ are not both observable. Define

$$Z_{ji} = \min(X_{ji}, C_{ji});$$

$$\varepsilon_{ji} = I_{\{X_{ji} \leq C_{ji}\}}, \quad j = 1,2; \; i = 1,2,\ldots,$$

where $I_A(x)$ is 1(0) if $x \in (\notin) A$. Note that ε corresponds to whether Z is an uncensored value ($\varepsilon = 1$) or censored value ($\varepsilon = 0$). It is assumed that $\underset{\sim}{Z}_i = (Z_{1i}, Z_{2i})$ and $\underset{\sim}{\varepsilon}_i = (\varepsilon_{1i}, \varepsilon_{2i})$ are observable. Let H denote the distribution function of $\underset{\sim}{Z}_i$. Define the bivariate survival function

$$\overline{F}(\underset{\sim}{t}) \equiv \overline{F}(t_1, t_2) \equiv P(X_1 > t_1, X_2 > t_2) \quad ,$$

and with abuse of notation, for $\underset{\sim}{t} = (t_1, t_2)$ define

$$F(\overline{t}_1, t_2) = P(X_1 > t_1, X_2 \leq t_2); \; F(t_1, \overline{t}_2) = P(X_1 \leq t_1, X_2 > t_2) \quad .$$

Similar functions can be defined for G and H. By independence of $\underset{\sim}{X}$ and $\underset{\sim}{C}$

(1) $$\overline{H}(t_1, t_2) = \overline{F}(t_1, t_2)\, \overline{G}(t_1, t_2)$$

for all t_1, t_2.

The hazard gradient approach of Marshall (1975) was employed by Campbell and Földes (1982) to estimate the distribution function as indicated below. The cumulative hazard function is given by

(2) $$R(\underset{\sim}{t}) = -\ell n\, \overline{F}(\underset{\sim}{t}) \quad .$$

Assuming R is absolutely continuous with partial derivatives that exist almost everywhere, let $\underset{\sim}{r}(\underset{\sim}{t})$ denote the gradient of $R(\underset{\sim}{t})$. Then $R(\underset{\sim}{t})$ can be represented as the path-independent integral of $\underset{\sim}{r}(\underset{\sim}{z})$ from (0,0) to $\underset{\sim}{t}$. In particular, for the linear path (0,0) to $(t_1,0)$ to (t_1,t_2) ,

$$R(\underset{\sim}{t}) = \int_0^{t_1} r_1(u,0)du + \int_0^{t_2} r_2(t_1,v)dv \quad .$$

Here $r_1(\underset{\sim}{s}) = \dfrac{\partial R(\underset{\sim}{s})}{\partial s_1}$, $r_2(\underset{\sim}{s}) = \dfrac{\partial R(\underset{\sim}{s})}{\partial s_2}$ for $\underset{\sim}{s} = (s_1,s_2)$; i.e.,

$$R(\underset{\sim}{t}) = \int_0^{t_1} (\bar{F}(u,0))^{-1} \, d_u F(u,\bar{0}) + \int_0^{t_2} (\bar{F}(t_1,v))^{-1} \, d_v F(\bar{t}_1,v) \ ,$$

where $d_u F(u,\bar{s})$ and $d_v F(\bar{s},v)$ denote Lebesque-Stieljes integration over u and v, respectively, with s fixed.

Define

$$K_1(\underset{\sim}{t}) = P(Z_1 \le t_1,\ Z_2 > t_2,\ \varepsilon_1 = 1) = \int_0^{t_1} \bar{G}(u,t_2) d_u F(u,\bar{t}_2) \ ;$$

(3)

$$K_2(\underset{\sim}{t}) = P(Z_1 > t_1,\ Z_2 \le t_2,\ \varepsilon_2 = 1) = \int_0^{t_2} \bar{G}(t_1,v) d_v F(\bar{t}_1,v) \quad .$$

Then

(4) $$R(\underset{\sim}{t}) = \int_0^{t_1} (\bar{H}(u,0))^{-1} \, d_u K_1(u,0) + \int_0^{t_2} (\bar{H}(t_1,v))^{-1} \, d_v K_2(t_1,v) \quad .$$

Estimate H, K_1 and K_2 by the empiricals

$$H_n(\underset{\sim}{t}) = \frac{1}{n} \sum_{i=1}^{n} I_{(Z_{1i} \le t_1,\ Z_{2i} \le t_2)} \ ;$$

$$K_{1n}(\underset{\sim}{t}) = \frac{1}{n} \sum_{i=1}^{n} \alpha_{1i}(\underset{\sim}{t}) \ ;$$

$$K_{2n}(\underset{\sim}{t}) = \frac{1}{n} \sum_{i=1}^{n} \alpha_{2i}(\underset{\sim}{t}) \ ;$$

where $\alpha_{1i}(\underset{\sim}{t}) = I_{(Z_{1i} \le t_1,\ Z_{2i} > t_2,\ \varepsilon_{1i} = 1)}$ and $\alpha_{2i}(\underset{\sim}{t}) = I_{(Z_{1i} > t_1,\ Z_{2i} \le t_2, \varepsilon_{2i} = 1)}$.

Then R(t) is estimated from $\{\underset{\sim}{Z}_i\}_{i=1}^n$ and $\{\underset{\sim}{\varepsilon}_i\}_{i=1}^n$ by

$$(5) \qquad R_n(\underset{\sim}{t}) = \int_0^{t_1} (\bar{H}_n(u^-,0))^{-1}\, d_u K_{1n}(u,0) + \int_0^{t_2} (\bar{H}_n(t_1,v^-))^{-1}\, d_v K_{2n}(t_1,v) ,$$

and $\bar{F}(\underset{\sim}{t})$ by

$$(6) \qquad \bar{F}_n(\underset{\sim}{t}) = \exp\{-R_n(\underset{\sim}{t})\} .$$

If F and G are continuous and if T_1 and T_2 are such that $\bar{H}(T_1,T_2) > 0$, Campbell and Földes (1982) proved

$$\sup_{\substack{0<t_1\le T_1 \\ 0<t_2\le T_2}} |\bar{F}_n(\underset{\sim}{t}) - \bar{F}(\underset{\sim}{t})| = 0\left(\sqrt{\frac{\ell n\ \ell n\ n}{n}}\right) \text{ a.s.}$$

3. <u>Weak Convergence of Empiricals in Two-Dimensional Time</u>

The study of the weak convergence of empirical processes in multi-dimensional time culminated in articles by Neuhaus (1971) and Straf (1972). The approach of Neuhaus (1971) is the reference for the topological discussion below.

For simplicity one can reduce the domain of the bivariate distribution function F from $[0,\infty) \times [0,\infty)$ to the unit square, $[0,1] \times [0,1]$ by the transformation $u_1 = F(t_1,\bar{0})$ and $u_2 = F(t_2|t_1) = P(X_2 \le t_2|X_1 \le t_1)$, as suggested in Durbin (1970). The approach of Neuhaus (1971) is to restrict the real-valued functions from the unit square. For the point $\underset{\sim}{t}=(t_1,t_2)$ inside the unit square, let Q_1, Q_2, Q_3, Q_4 denote the four open quadrants in the square determined by $\underset{\sim}{t}$, where Q_1 is the upper right quadrant. The space D_2 is the set of all real functions from the unit square such that if $\{\underset{\sim}{t}_n\}$ denotes a sequence in Q_i such that $\lim_{n\to\infty} \underset{\sim}{t}_n = \underset{\sim}{t}$ then $\lim_{n\to\infty} f(\underset{\sim}{t}_n)$ exists for $i = 1,2,3,4$ and for $i = 1$ its limiting value is $f(\underset{\sim}{t})$. Let Λ denote the class of all continuous functions from $[0,1]$ onto itself. Let $\underset{\sim}{\lambda} = (\lambda_1, \lambda_2) \in \Lambda \times \Lambda$ and $|\underset{\sim}{t}|$ denote Euclidean distance in the plane. Define the metric d (which can be thought of as an extension of the one-dimensional Skorohod metric) for f,g in D_2 as

$$d(f,g) = \inf_{\varepsilon>0}\{\varepsilon: \text{there exists } \underset{\sim}{\lambda} = \underset{\sim}{\lambda}_\varepsilon \text{ with } \sup_{\underset{\sim}{t}}|\underset{\sim}{\lambda}(\underset{\sim}{t}) - \underset{\sim}{t}| \leq \varepsilon$$
$$\text{and } \sup_{\underset{\sim}{t}}|f(\underset{\sim}{t}) - g(\underset{\sim}{\lambda}(\underset{\sim}{t}))| \leq \varepsilon\} .$$

Then (D_2, d) is a separable complete metric space, unlike the space of discontinuous functions in the unit square with the metric d or the sup metric. Therefore, the Prohorov development of weak convergence is applicable.

Let

$$U_n(\underset{\sim}{t}) = \sqrt{n}\,(\bar{H}_n(t_1,t_2) - \bar{H}(t_1,t_2)) ;$$

(7)

$$V_{jn}(\underset{\sim}{t}) = \sqrt{n}\,(K_{jn}(\underset{\sim}{t}) - K_j(\underset{\sim}{t})) , \quad j = 1,2 .$$

LEMMA :

Let T_1, T_2 be such that $\bar{H}(T_1,T_2) > 0$. Then as $n \to \infty$ $(U_n(\underset{\sim}{t}), V_{1n}(\underset{\sim}{t}), V_{2n}(\underset{\sim}{t}))$ converges weakly to a trivariate, two-dimensional-time Gaussian process $(U(\underset{\sim}{t}), V_1(\underset{\sim}{t}), V_2(\underset{\sim}{t}))$ with mean (0,0,0) and covariance structure given below, where $a \wedge b = \min(a,b)$, $a \vee b = \max(a,b)$, and $\underset{\sim}{s} = (s_1, s_2)$:

$$\mathrm{Cov}(U(\underset{\sim}{s}),U(\underset{\sim}{t})) = \bar{H}(s_1 \vee t_1,\ s_2 \vee t_2) - \bar{H}(s_1,s_2)\ \bar{H}(t_1,t_2)\ ;$$

$$\mathrm{Cov}(V_1(\underset{\sim}{s}),V_1(\underset{\sim}{t})) = K_1(s_1 \wedge t_1,\ s_2 \vee t_2) - K_1(s_1,s_2)\ K_1(t_1,t_2)\ ;$$

$$\mathrm{Cov}(V_2(\underset{\sim}{s}),V_2(\underset{\sim}{t})) = K_2(s_1 \vee t_1,\ s_2 \wedge t_2) - K_2(s_1,s_2)\ K_2(t_1,t_2)\ ;$$

$$\mathrm{Cov}(U(\underset{\sim}{s}),V_1(\underset{\sim}{t})) = \begin{cases} -\bar{H}(s_1,s_2)\ K_1(t_1,t_2) & \text{if } s_1 \geq t_1 \\ K_1(t_1,s_2 \vee t_2) - K(s_1,s_2 \vee t_2) - \bar{H}(s_1,s_2)\ K_1(t_1,t_2) & \text{if } s_1 < t_1; \end{cases}$$

$$\mathrm{Cov}(U(\underset{\sim}{s}),V_2(\underset{\sim}{t})) = \begin{cases} -\bar{H}(s_1,s_2)\ K_2(t_1,t_2) & \text{if } s_2 \geq t_2 \\ K_2(s_1 \vee t_1,t_2) - K_2(s_1 \vee t_1,s_2) - \bar{H}(s_1,s_2)\ K_2(t_1,t_2) & \text{if } s_2 < t_2; \end{cases}$$

$$\mathrm{Cov}(V_1(\underset{\sim}{s}),V_2(\underset{\sim}{t})) = \begin{cases} -K_1(s_1,s_2)\ K_2(t_1,t_2) & \text{if } s_1 < t_1 \text{ or } s_2 > t_2 \\ J(s_1,t_2) + J(t_1,s_2) - J(s_1,s_2) - J(t_1,t_2) & \\ \qquad - K_1(s_1,s_2)\ K_2(t_1,t_2) & \text{if } s_1 \geq t_1,\ s_2 \leq t_2\ , \end{cases}$$

where $J(t_1,t_2) = P(Z_1 \leq t_1,\ Z_2 \leq t_2,\ \varepsilon_1 = 1,\ \varepsilon_2 = 1)$.

PROOF:

The finite dimensional distributions converge to multivariate normal distributions by an application of the multivariate central limit theorem to the three-dimensional variables

$$\Big(I_{(Z_{1i} > t_1,\ Z_{2i} > t_2)} - \bar{H}(t_1,t_2)\ ,$$

$$I_{(Z_{1i} \leq t_1,\ Z_{2i} > t_2,\ \varepsilon_{1i} = 1)} - K_1(\underset{\sim}{t})\ ,\quad I_{(Z_{1i} > t_1,\ Z_{2i} \leq t_2,\ \varepsilon_{2i} = 1)} - K_2(\underset{\sim}{t}) \Big).$$

A simple calculation on these indicator variables yields the covariance structure of the lemma. In order to prove tightness in three dimensions, the tightness result of Neuhaus (1971, pp. 1292-5) for a one-dimensional empirical function of a multidimensional time parameter is applied for U_n, V_{1n} and V_{2n}

separately. Then tightness of the distributions of (U_n, V_{1n}, V_{2n}) follows immediately.

4. Main Theorems

The theorems in this section proceed in a fashion similar to the proof of weak convergence in one dimension in Breslow and Crowley (1974). The random variables (U_n, V_{1n}, V_{2n}) and (U, V_1, V_2) can be replaced by random variables with the same finite dimensional distribution but which also satisfy the condition that $d((U_n, V_{1n}, V_{2n}), (U, V_1, V_2))$ converges to zero almost surely, where d also represents the extension to $D_2 \times D_2 \times D_2$ of the metric d on D_2.

THEOREM 1:

If F and G are continuous bivariate distribution functions and if $T_1, T_2 < \infty$ are such that $\bar{H}(T_1, T_2) > 0$, then for $\underset{\sim}{t} = (t_1, t_2)$ with $0 < t_1 < T_1$, $0 < t_2 < T_2$,

$$\sqrt{n}\,(R_n(\underset{\sim}{t}) - R(\underset{\sim}{t}))$$

converges weakly to a two-dimensional-time Gaussian process $W(\underset{\sim}{t})$ given by:

$$W(\underset{\sim}{t}) = A_1(\underset{\sim}{t}) + B_1(\underset{\sim}{t}) + A_2(\underset{\sim}{t}) + B_2(\underset{\sim}{t}) ,$$

where

$$(8)\quad \begin{aligned} A_1(\underset{\sim}{t}) &= -\int_0^{t_1} (\bar{H}(u,0))^{-2}\, U(u,0)\, d_u K_1(u,0) \\ B_1(\underset{\sim}{t}) &= (\bar{H}(t_1,0))^{-1}\, V_1(t_1,0) - \int_0^{t_1} (\bar{H}(u,0))^{-2}\, V_1(u,0)\, d_u H(u,\bar{0}) \\ A_2(\underset{\sim}{t}) &= -\int_0^{t_2} (\bar{H}(t_1,v))^{-2}\, U(t_1,v)\, d_v K_2(t_1,v) \\ B_2(\underset{\sim}{t}) &= (\bar{H}(t_1,t_2))^{-2}\, V_2(t_1,t_2) - \int_0^{t_2} (\bar{H}(t_1,v))^{-2}\, V_2(t_1,v)\, d_v H(\bar{t}_1,v) . \end{aligned}$$

PROOF:

From equations (4) and (5)

$$\sqrt{n}\,(R_n(\underset{\sim}{t}) - R(t)) = \sqrt{n}\,[\int_0^{t_1} (\bar{H}_n(u^-,0))^{-1}\, d_u K_{1n}(u,0) - \int_0^{t_1} (\bar{H}(u,0))^{-1}\, d_u K_1(u,0)$$

$$+ \int_0^{t_2} (\bar{H}_n(t_1,v^-))^{-1}\, d_v K_{2n}(t_1,v) - \int_0^{t_2} (\bar{H}(t_1,v))^{-1}\, d_v K_2(t_1,v)]\,.$$

Now

$$\sqrt{n}\,[\int_0^{t_1} (\bar{H}_n(u^-,0))^{-1}\, d_u K_{1n}(u,0) - \int_0^{t_1} (\bar{H}(u,0))^{-1}\, d_u K_1(u,0)]$$

$$= \sqrt{n}\left\{\int_0^{t_1} (\bar{H}(u,0))^{-2}\, [\bar{H}(u,0) - H_n(u^-,0)]\, d_u K_1(u,0)\right.$$

$$+ \int_0^{t_1} (\bar{H}(u,0))^{-1}\, d_u (K_{1n}(u,0) - K_1(u,0))$$

$$+ \int_0^{t_1} [\bar{H}(u,0) - \bar{H}_n(u^-,0)]\left[\frac{1}{\bar{H}(u,0)\,\bar{H}_n(u^-,0)} - \frac{1}{\bar{H}^2(u,0)}\right] d_u K_1(u,0)$$

$$\left. + \int_0^{t_1} \left[\frac{1}{\bar{H}_n(u^-,0)} - \frac{1}{\bar{H}(u,0)}\right] d_u (K_{1n}(u,0) - K_1(u,0))\right\}$$

$$= A_{1n}(\underset{\sim}{t}) + B_{1n}(\underset{\sim}{t}) + E_{1n}(\underset{\sim}{t}) + E_{1n}^*(\underset{\sim}{t})\ ,$$

where

$$A_{1n}(\underset{\sim}{t}) = -\int_0^{t_1} (\bar{H}(u,0))^{-2}\, U_n(u,0)\, d_u K_1(u,0)$$

$$B_{1n}(\underset{\sim}{t}) = (\bar{H}(t_1,0))^{-1}\, V_{1n}(t_1,0) - \int_0^{t_1} (\bar{H}(u,0))^{-2}\, V_{1n}(u,0)\, d_u H(u,\bar{0})$$

$$E_{1n}(\underset{\sim}{t}) = \frac{1}{\sqrt{n}}\int_0^{t_1} \frac{U_n^2(t_1,0)}{\bar{H}_n(u^-,0)\,\bar{H}^2(u,0)}\, d_u K_1(u,0)$$

$$E_{1n}^*(t) = -\int_0^{t_1} \frac{U_n(u,0)}{\bar{H}_n(u^-,0)\,\bar{H}(u,0)}\, d_u (K_{1n}(u,0) - K_1(u,0))\ .$$

In a similar manner,

$$\sqrt{n}\left[\int_0^{t_2} (\bar{H}_n(t_1,v^-))^{-1}\, d_v K_{2n}(t_1,v) - \int_0^{t_2} (\bar{H}(t_1,v))^{-1}\, d_v K_2(t_1,v)\right]$$

$$= A_{2n}(\underset{\sim}{t}) + B_{2n}(\underset{\sim}{t}) + E_{2n}(\underset{\sim}{t}) + E^*_{2n}(\underset{\sim}{t}) \;,$$

where

$$A_{2n}(\underset{\sim}{t}) = -\int_0^{t_2} (\bar{H}(t_1,v))^{-2}\, U_n(t_1,v)\, d_v K_2(t_1,v)$$

$$B_{2n}(\underset{\sim}{t}) = (\bar{H}(t_1,t_2))^{-1}\, V_{2n}(t_1,t_2) - \int_0^{t_2} (\bar{H}(t_1,v))^{-2}\, V_{2n}(t_1,v)\, d\bar{H}(t_1,v)$$

$$E_{2n}(\underset{\sim}{t}) = \frac{1}{\sqrt{n}} \int_0^{t_2} \frac{U_n^2(t_1,v)}{\bar{H}_n(t_1,v^-)\,\bar{H}^2(t_1,v)}\, d_v K_2(t_1,v)$$

$$E^*_{2n}(\underset{\sim}{t}) = -\int_0^{t_2} \frac{U_n(t_1,v)}{\bar{H}_n(t_1,v^-)\,\bar{H}(t_1,v)}\, d_v (K_{2n}(t_1,v) - K_2(t_1,v)) \;.$$

Now as n tends to infinity, $E_{jn}(\underset{\sim}{t})$ and $E^*_{jn}(\underset{\sim}{t})$ converge in probability to zero in the supremum metric by an argument similar to that of Breslow and Crowley (1974) and hence converge in probability to zero in the metric d, for $j = 1,2$. Further, $A_{jn}(\underset{\sim}{t})$ converges almost surely to $A_j(\underset{\sim}{t})$ and $B_{jn}(\underset{\sim}{t})$ to $B_j(\underset{\sim}{t})$ in the sup metric and hence in d, for $j = 1,2$. Therefore, $\sqrt{n}(R_n(\underset{\sim}{t}) - R(\underset{\sim}{t}))$ converges weakly to $W(\underset{\sim}{t})$. That $W(\underset{\sim}{t})$ is a Gaussian process with mean 0 follows immediately from the Lemma. The covariance structure of $W(\underset{\sim}{t})$ can be calculated from (8) and from the covariance structure of (U,V_1,V_2).

THEOREM 2:

If F and G are continuous bivariate distribution functions and if $T_1,T_2 < \infty$ are such that $\bar{H}(T_1,T_2) < \infty$, then for $0 < t_1 < T_1$, $0 < t_2 < T_2$

$$\sqrt{n}\ (\bar{F}_n(\underset{\sim}{t}) - \bar{F}(\underset{\sim}{t}))$$

converges weakly to a two-dimensional-time Gaussain process $W^*(\underset{\sim}{t})$ which has mean 0 and covariance

$$\mathrm{Cov}(W^*(s_1,s_2),\ W^*(t_1,t_2)) = \bar{F}(s_1,s_2)\ \bar{F}(t_1,t_2)\ \mathrm{Cov}(W(s_1,s_2),\ W(t_1,t_2))\ .$$

PROOF:

By (6)

$$\sqrt{n}\ (\bar{F}_n(\underset{\sim}{t}) - \bar{F}(\underset{\sim}{t})) = \sqrt{n}\,(e^{-R_n(\underset{\sim}{t})} - e^{-R(\underset{\sim}{t})})\ .$$

A Taylor's series expansion yields

$$(9)\qquad \sqrt{n}\,(\bar{F}_n(\underset{\sim}{t}) - \bar{F}(\underset{\sim}{t})) = -e^{-R(\underset{\sim}{t})}\ \sqrt{n}\,(R_n(\underset{\sim}{t}) - R(\underset{\sim}{t})) - e^{-R_n^*(\underset{\sim}{t})}\ \sqrt{n}\,(R_n(\underset{\sim}{t}) - R(\underset{\sim}{t}))^2\ ,$$

where $\sup|R_n^* - R| \leq \sup|R_n - R|$. The second term on the right of (9) converges to zero in probability in the sup norm and hence in d. Thus, $\sqrt{n}\,(\bar{F}_n(\underset{\sim}{t}) - \bar{F}(\underset{\sim}{t}))$ converges weakly to $-\bar{F}(\underset{\sim}{t})\ W(\underset{\sim}{t})$ which is a Gaussian process with mean 0 and and desired covariance.

5. Other Estimators

Campbell and Földes (1982) introduced another path-dependent estimator. For $N(\underset{\sim}{t}) = n\ \bar{H}_n(\underset{\sim}{t})$, define the estimator $S_n(\underset{\sim}{t})$ of $\bar{F}(\underset{\sim}{t})$:

$$S_n(\underset{\sim}{t}) = \prod_{i=1}^{n} \left(\frac{N(Z_{1i},0)}{N(Z_{1i},0)+1}\right)^{\alpha_{1i}(t_1,0)} \prod_{i=1}^{n} \left(\frac{N(t_1,Z_{2i})}{N(t_1,Z_{2i})+1}\right)^{\alpha_{2i}(t_1,t_2)}\ ,$$

provided $N(\underset{\sim}{t}) > 0$. This is the product of two one-dimensional product-limit estimators, one a marginal estimator for the first coordinate, and the second a conditional one on the second coordinate given $Z_1 > t_1$. It has been proved that for $T_1, T_2 < \infty$ such that $\overline{H}(T_1, T_2) > 0$

$$\sup_{\substack{0<t_1\leq T_1 \\ 0<t_2\leq T_2}} |\overline{F}_n(\underset{\sim}{t}) - S_n(\underset{\sim}{t})| = o\left(\frac{1}{n}\right) \text{ a.s.}$$

Therefore $S_n(\underset{\sim}{t})$ inherits strong uniform almost sure consistency as well as weak convergence from $\overline{F}_n(\underset{\sim}{t})$.

The estimators $\overline{F}_n(\underset{\sim}{t})$ and $S_n(\underset{\sim}{t})$ depend on the path from (0,0) to $(t_1,0)$ to (t_1,t_2). Since $R(\underset{\sim}{t})$ and $F(\underset{\sim}{t})$ are path independent, it is possible to have also developed estimators for the linear path from (0,0) to $(0,t_2)$ to (t_1,t_2). In general these estimators differ from $\overline{F}_n(\underset{\sim}{t})$ and $S_n(\underset{\sim}{t})$. However, the strong consistency and weak convergence results follow in the same way. The covariance structure of the limiting Gaussian process does depend on the path.

The extension of these estimators and their asymptotic properites from two-dimensional time to k(>2)-dimensional time is straightforward. In k-dimensions there are k! piecewise linear paths similar to the two mentioned above. Strong uniform almost consistency follows readily. It is convenient to use the definition of D_k in Neuhaus (1971) to develop the weak convergence results.

The drawback of the estimators $\overline{F}_n$ and S_n is that the estimators are not necessarily survival distribution functions. Although monotonicity is assured along the path of definition, it is not guaranteed along other ever-increasing paths nor is it the case that non-negative mass is assigned to rectangles. The fact that the estimator is uniformly strongly consistent for the properly behaved function $\overline{F}$ minimizes this problem for large samples. In that S_n can be thought of as a generalized maximum likelihood estimator along the designated path, one could obtain such a generalized maximum likelihood estimator subject to the constraint that the estimator is a distribution function. It is conjectured that these asymptotic results will not be changed for such an estimator.

REFERENCES

Breslow, N. and Crowley, J. (1974). A large sample study of the life table and product limit estimates under random censorship. Annals of Statistics 2, 437-453.

Campbell, G. (1981). Nonparametric bivariate estimation with randomly censored data. Biometrika 68, 417-422.

Campbell, G. and Földes, A. (1982). Large-sample properites of nonparametric bivariate estimators with censored data. Proceedings of the International Colloquium on Nonparametric Statistical Inference, 1980, North Holland. To appear.

Durbin, J. (1970). Asymptotic distributions of some statistics based on the bivariate sample distribution function. In Nonparametric Techniques in Statistical Inference, Ed. M.L. Puri. Cambridge University Press, pp. 435-449.

Földes, A. and Rejtö, L. (1981). Strong uniform consistency for nonparametric survival curve estimators from randomly censored data. Annals of Statistics 9, 122-129.

Kaplan, E.L. and Meier, P. (1958). Nonparametric estimation from incomplete observations. Journal of the American Statistical Association 53, 457-481.

Marshall, A.W. (1975). Some comments on the hazard gradient. Stochastic Processes and Their Applications 3, 293-300.

Neuhaus, G. (1971). On weak convergence of stochastic processes with multidimensional time parameter. Annals of Mathematical Statistics 42, 1285-1295.

Straf, M.L. (1972). Weak convergence of stochastic processes with several parameters. Proceedings of the Sixth Berkeley Symposium on Mathematical Statistics and Probability, V.II, University of California Press, Berkeley, California, pp 187-221.

GROUP SEQUENTIAL METHODS FOR SURVIVAL ANALYSIS WITH STAGGERED ENTRY

Anastasios A. Tsiatis

Harvard University School of Public Health and Sidney Farber
Cancer Institute Boston, Massachusetts

1. Introduction

In many clinical trials, especially in the study of chronic disease, we are often interested in the comparison of time to failure among different treatment groups. Many procedures have been developed in the past decade to analyze such failure time data, the most popular being the logrank test (Mantel, 1966; Peto and Peto, 1972) and modifications of the Wilcoxon test (Gilbert, 1962; Gehan, 1965; Breslow, 1970; Peto and Peto, 1972; Prentice, 1978).

Typically in such a trial, patients enter the study serially, are then assigned according to some random mechanism to different treatment arms and are followed until they either fail or the study is terminated. Ordinarily, these studies are designed so that after sufficient amount of patient accural and follow-up time a single terminal analysis will be made to test whether the failure time distribution is the same among the different treatment groups. In practice, however, as well as for ethical considerations, the data are monitored periodically and if sufficient differences are found between the treatment groups, a decision might be made to stop the study early. It is, therefore, very important to study the sequential properties of the tests used in survival analysis in order that correct and efficient methods be employed in monitoring the data.

Breslow (1969) and Breslow and Haug (1972) provide sequential methods of comparing exponential distributions. However, most of the recent work examines the more commonly used nonparametric statistics. Jones and Whitehead (1979) have looked at the sequential logrank and modified Wilcoxon tests. Nagelkerke and Hart (1980) also indicate how to extend the sequential probability ratio test using partial likelihoods. In a more rigorous fashion Tsiatis (1981) has derived the group sequential distribution of the logrank score and Slud and Wei (1982) that of the modified Wilcoxon score.

Most of the commonly used nonparametric statistics are special cases of a general class characterized by Tarone and Ware (1977) and Prentice and Marek (1979). In this paper, the asymptotic joint distribution of the sequentially computed test statistics, within this general class of nonparametric tests, will be derived.

These results will prove useful in constructing group sequential procedures.

2. Notation and Formulae

Let the positive random variables X,Y denote failure time and time of entry into study, respectively. Also let the positive random variable W denote time to censoring such as loss to follow-up. We wish to test the null hypothesis, H_0, that the hazard rate for failure is not related to a covariate Z. That is

$$H_0 : \lambda(x|z) = \lambda(x) ,$$

for all $x \geq 0$, where $\lambda(x|z)$ denotes the conditional hazard rate at time x given that the covariate Z is equal to z.

The time of entry into study, Y , and time to censoring, W , will have distributions $H(y|z) = P(Y \leq y|Z = z)$ and $G(w|z) = P(W \leq w|Z = z)$, respectively, which may depend on the covariate Z . We will denote the survival distribution $1 - G(w|z)$ as $\bar{G}(w|z)$. Assume that, given the covariate Z, the random variables

X,Y,W are independent.

If the data were examined at time t, the following variables could be observed; time to failure or censoring $X(t) = \max\{\min(X, t-Y, W), 0\}$, and an indicator variable for failure $\Delta(t) = 1$ if $X < \min(t-Y, W)$, $\Delta(t) = 0$ otherwise. At time t, the data can be represented as n identically and independently distributed random vectors $\{X_i(t), \Delta_i(t), Z_i\}$ for $i = 1, \ldots, n$.

The class of tests for testing H_0 will be similar to those of Tarone and Ware (1977) and Prentice and Marek (1979). Using the notation of this paper, we define a class of tests Θ, characterized by statistics of the form

$$S_n(t) = \sum_{i=1}^{n} \hat{Q}(t, X_i(t))\, \Delta_i(t)\, \{Z_i - \sum_{j \in R(t, X_i(t))} Z_j / n(t, X_i(t))\} ,$$

where R(t,x) denotes the risk set at time x if the data were observed at real time t, $x \le t$. That is, R(t,x) denotes the set of indices $\{j = 1, \ldots, n\}$ such that $\{X_j(t) \ge x\}$. Letting I(A) denote the indicator function of the event A, then $n(t,x) = \sum_{j=1}^{n} I(X_j(t) \ge x)$. For fixed t, the random function $\hat{Q}(t,x)$, $x \le t$, is assumed to converge in probability in sup norm to a function Q(t,x).

The random function $\hat{Q}(t,x)$ corresponds to the weighting functions W_i described by Tarone and Ware (1977) and Prentice and Marek (1979). In particular, the most widely used nonparametric tests can be represented as follows:

Example 1. For the logrank test, $\hat{Q}(t,x) = Q(t,x) = 1$ for all $t > 0$, $0 \le x \le t$.

Example 2. The weighting function for the modified-Wilcoxon test is given by $\hat{Q}(t,x) = n(t,x)/n$, which converges in probability to the function $Q(t,x) = P(X(t) \ge x)$. Due to the independence of X,Y,W given Z, this can be expressed, under H_0, as

$$P(X(t) \geq x) = E\{P(X(t) \geq x) \mid Z\}$$

(1) $$= E\{P(X \geq x,\ t - Y \geq x,\ W \geq x) \mid Z\}$$

$$= \exp\{-\Lambda(x)\}\ E\{H(t - x) \mid Z\}\ E\{\bar{G}(x) \mid Z\} \quad .$$

Example 3. The weighting function for Prentice's (1978) generalization of the Wilcoxon test can be expressed as

$$\hat{Q}(t,x) = KM(x) = \prod_{i=1}^{n} [n(t,X_i(t))/\{n(t,X_i(t)) + 1\}]^{N_i(t,x)} \quad ,$$

where

(2) $$N_i(t,x) = I(X_i(t) \leq x,\ \Delta_i(t) = 1) \quad .$$

We note that, under H_0, $Q(t,x)$ is approximately the same as the Kaplan and Meier (1958) estimate of the survival distribution. Consequently, $Q(t,x) = \exp\{-\Lambda(x)\}$.

Example 4. The tests based on the score statistics, called G^ρ tests, proposed by Harrington and Fleming (1981) have weighting functions $\hat{Q}(t,x) = \{KM(x)\}^\rho$, and hence $Q(t,x) = [\exp\{-\Lambda(x)\}]^\rho$.

3. Asymptotic Joint Distribution of the Statistic

The key to deriving the joint distribution of the statistic $S_n(t)$ over time is to approximate it by a sum of i.i.d. random variables. All subsequent results are assumed under the null hypothesis.

We first note that the general statistic can be written as

(3) $$S_n(t) = \sum_{i=1}^{n} \int_0^t dN_i(t,x)\ \hat{Q}(t,x)\ \{Z_i - \sum_{j \in R(t,x)} Z_j/n(t,x)\} \quad ,$$

where $N_i(t,x)$ is given by (2).

For fixed t, $N_i(t,x)$ is a counting process as a function of x with intensity process given by $\lambda(x)\ I(X_i(t) \geq x)$. Therefore, using the results of Aalen (1977, 1978), the process

$$M_i(t,x) = N_i(t,x) - \int_0^x \lambda(u)\ I(X_i(t) \geq u)du$$

is a martingale. The statistic (3) can then be shown to be equal to

$$S_n(t) = \sum_{i=1}^{n} \int_0^t \hat{Q}(t,x)\ dM_i(t,x)\ \{Z_i - \sum_{j\varepsilon R(t,x)} Z_j/n(t,x)\} \quad .$$

Because of the complex relationship between t and x, the general martingale approach of Aalen is not directly applicable in characterizing the asymptotic joint distribution of the process $S_n(t)$. However, by noting that

$$\sum_{j\varepsilon R(t,x)} Z_j/n(t,x)$$

converges in probability to

(4) $$\mu(t,x) = [E\{ZH(t-x\,|Z)\ \bar{G}(x|Z)\}] / [E\{H(t-x|Z)\ \bar{G}(x|Z)\}] \quad ,$$

and $\hat{Q}(t,x)$ converges in probability to $Q(t,x)$, then the statistic $n^{-\frac{1}{2}}S_n(t)$ can be approximated by $n^{-\frac{1}{2}}\bar{S}_n(t)$, where

(5) $$\bar{S}_n(t) = \sum_{i=1}^{n} \int_0^t Q(t,x)\ dM_i(t,x)\ \{Z_i - \mu(t,x)\} \quad .$$

The difference between $n^{-\frac{1}{2}}S_n(t)$ and $n^{-\frac{1}{2}}\bar{S}_n(t)$ can be shown to be asymptotically negligible by using results similar to Tsiatis (1981b, Lemma 3.1).

The approximate statistic $\bar{S}_n(t)$ can be written as

$$\bar{S}_n(t) = \sum_{i=1}^{n} [\Delta_i(t)\, Q(t,X_i(t))\, \{Z_i - \mu(t,X_i(t))\}$$

(6)

$$- \int_0^{X_i(t)} Q(t,x)\, \{Z_i - \mu(t,x)\}\, \lambda(x)dx] \quad .$$

Although (6) is complex it is nonetheless a sum of identically and independently distributed random variables and the asymptotic distribution can be obtained by application of the central limit theorem. We are now in a position to prove the following fundamental theorem.

THEOREM 3.1. Defining the statistics in Θ by

$$S_n^{(1)}(t) = \sum_{i=1}^{n} \hat{Q}_1(t,X_i(t))\, \Delta_i(t)\, \{Z_i - \sum_{j\varepsilon R(t,X_i(t))} Z_j/n(t,X_i(t))\} ,$$

and $$S_n^{(2)}(t') = \sum_{i=1}^{n} \hat{Q}_2(t',X_i(t'))\, \Delta_i(t')\{Z_i - \sum_{j\varepsilon R(t',X_i(t'))} Z_j/n(t',X_i(t'))\} ,$$

where $t' \geq t$; then the random vector $n^{-\frac{1}{2}}\{S_n^{(1)}(t),\, S_n^{(2)}(t')\}$ converges in distribution to a bivariate normal distribution with mean zero and covariance matrix

$$\Omega = \begin{bmatrix} \sigma_{11} & \sigma_{12} \\ \sigma_{12} & \sigma_{22} \end{bmatrix}$$

where

$$\sigma_{11} = \int_0^t Q_1^2(t,x)\, \psi(t,x)\, \lambda(x)dx ,$$

(7) $$\sigma_{12} = \int_0^t Q_1(t,x)\, Q_2(t',x)\, \psi(t,x)\, \lambda(x)dx ,$$

$$\sigma_{22} = \int_0^{t'} Q_2^2(t',x)\, \psi(t',x)\, \lambda(x)dx ,$$

and $$\psi(t,x) = E[\{Z - \mu(t,x)\}^2\, H(t-x|Z)\, \bar{G}(x|Z)]\, \exp\{-\Lambda(x)\} .$$

PROOF. Because of the asymptotic approximation of

$$n^{-\frac{1}{2}}\{S_n^{(1)}(t),\ S_n^{(2)}(t')\} \text{ for } t' \geq t ,$$

to

$$n^{-\frac{1}{2}}\{\bar{S}_n^{(1)}(t),\ \bar{S}_n^{(2)}(t')\} ,$$

it suffices to find the asymptotic joint distribution of the latter. Since this is a normalized sum of i.i.d. random vectors then a routine application of the multivariate central limit theorem together with the calculation of the moments given in the appendix of Tsiatis (1982) will yield the desired results.

It can also be shown by using the results of Tsiatis (1981a) that the covariance matrix Ω can be consistently estimated by replacing the quantities in (7) by their appropriate empirical estimates. In particular, the estimate of σ_{12} is given by

$$\hat{\sigma}_{12} = \int_0^t \hat{Q}_1(t,x)\ \hat{Q}_2(t',x)\ \hat{E}[\{Z - \hat{\mu}(t,x)\}^2\ I(X(t) \geq x)]\ d\hat{\Lambda}(x) ,$$

where

$$\hat{E}[\{Z - \hat{\mu}(t,x)\}^2\ I(X_i(t) \geq x)] = \sum_{j \in R(t,x)} \{Z_j - \hat{\mu}(t,x)\}^2/n ,$$

and $\hat{\Lambda}(x)$ is the estimate of the cumulative hazard function given by Nelson (1969), namely

$$\hat{\Lambda}(x) = \int_0^x dN_i(t,u)/n(t,u) .$$

COMMENT: The assumptions in this paper will apply to a completely randomized clinical trial but will not reflect a study with dynamic treatment assignment. However, with some modifications to the proofs we can allow the covariates $Z_1,\ldots,Z_n$ to be fixed values. Replacing $\mu(t,x)$ and $\psi(t,x)$ given in (4) and (7) by the following quantities

$$\mu^n(t,x) = \sum_{i=1}^{n} Z_i H(t-x|Z_i)\, \bar{G}(x|Z_i) \Big/ \sum_{i=1}^{n} H(t-x|Z_i)\, \bar{G}(x|Z_i)$$

and

$$\psi^n(t,x) = \sum_{i=1}^{n} \{Z_i - \mu^n(t,x)\}^2\, H(t-x|Z_i)\, \bar{G}(x|Z_i)\, \exp\{-\Lambda(x)\} / n \quad ,$$

and by assuming that $\mu^n(t,x)$ and $\psi^n(t,x)$ converge to $\mu^*(t,x)$ and $\psi^*(t,x)$, as would be the case for dynamic treatment assignment, then the statistic $S_n(t)$ can be approximated by

$$\bar{S}_n(t) = \sum_{i=1}^{n} \int_0^t Q(t,x)\, dM_i(t,x)\, \{Z_i - \mu^n(t,x)\} \quad .$$

We can then prove the results of Theorem 3.1 for fixed covariates by applying a central limit theorem for independent but not identically distributed random variable to $n^{-\frac{1}{2}}\{\bar{S}_n^{(1)}(t),\ \bar{S}_n^{(2)}(t')\}$. In this particular case the asymptotic covariance matrix is given by σ_{ij}^* which is obtained by substituting $\psi^*(t,x)$ for $\psi(t,x)$ in formula (7). However, the consistent estimate for σ_{ij}^* is the same as $\hat{\sigma}_{ij}$ given above and hence all applications to group sequential tests given in this paper would be identical.

4. Concluding Remarks

The asymptotic convergence of the joint distribution of test statistics within the class Θ, evaluated at time points $t_1 < t_2 < \cdots < t_k$, to a multivariate normal with mean zero, and covariance matrix that can be estimated, will enable us to construct group sequential tests at those time points by using methods described by Slud and Wei (1982, Section 4). In particular, we shall consider group sequential tests that will reject the null hypothesis if

$$|S_n(t_1)| \geq d_1$$

or

$$|S_n(t_1)| < d_1, |S_n(t_2)| \geq d_2$$

or

$$\vdots$$

$$|S_n(t_1)| < d_1, \ldots, |S_n(t_{k-1})| < d_{k-1}, |S_n(t_k)| \geq d_k \quad ,$$

otherwise we accept H_0. The boundary values $d_1, \ldots, d_k$ are derived by choosing $\alpha_1, \ldots, \alpha_k$ so that $\alpha_1 + \cdots + \alpha_k = \alpha$ and then recursively solving the following equations:

$$P(|U_1| \geq d_1) = \alpha_1$$

where $U_1 \sim N(0, \hat{\sigma}_{11})$

$$P(|U_1| < d_1, |U_2| \geq d_2) = \alpha_2$$

where $(U_1, U_2) \sim N\left(\underset{\sim}{0}, \begin{pmatrix} \hat{\sigma}_{11} & \hat{\sigma}_{12} \\ \hat{\sigma}_{12} & \hat{\sigma}_{22} \end{pmatrix}\right)$,

etc.

This method will guarantee an overall level of significance equal to some prespecified α. In order to apply these methods in practice, however, numerical methods for calculating multivariate normal integrals have to be used. Currently such methods are very inefficient. However, in many cases, as we shall indicate later, the score statistic has uncorrelated increments and for such instances recursive integration formulas given by Armitage, McPherson and Rowe (1969) can be used to solve the above equations.

Using the results of Theorem 3.1, we have the flexibility of constructing the group sequential test by using different test statistics within the class Θ at any of the time points $t_1,\ldots,t_k$. Since different tests are more sensitive to different types of alternatives this might allow us to construct more robust group sequential procedures.

We also note that if the weighting function $\hat{Q}(t,x)$ converges to a function Q(x), independent of t, then the score function $S_n(t)$ has asymptotically independent increments. This follows the fact that σ_{12} is equal to σ_{11}.

Therefore, the statistic for the sequentially computed logrank test, Prentice's generalization of the Wilcoxon test, and the G^{ρ} tests, described by Examples 1, 3, and 4, respectively, have independent increments. However, the modified-Wilcoxon test of Example 2 would not have this property unless all patients entered at once into the study. This contradicts Jones and Whitehead (1979) but supports the results of Slud and Wei (1982).

The results of Theorem 3.1 can be used to find the asymptotic joint distribution of any finite number of test statistics within the class θ. Therefore, the asymptotic joint distribution of the logrank test and modified-Wilcoxon test derived by Tarone (1981) would be a simple consequence of Theorem 3.1.

ACKNOWLEDGEMENT

This investigation was supported by grant NO. CA-23415, awarded by the National Cancer Institute, DHHS.

REFERENCES

Aalen, O.O. (1977). Weak convergence of stochastic integrals related to counting processes. Z. Wahrscheinlichkeitstheorie verw. Geb. **38**, 261-277.

Aalen, O.O. (1978). Nonparametric inference for a family of counting processes. Annals of Statistics **6**, 701-726.

Armitage, P., McPherson, C.K. and Rowe, B.C. (1969). Repeated significance tests on accumulating data. Journal of the Royal Statistical Society A **132**, 235-244.

Breslow, N. (1969). On large sample sequential analysis with applications to survivorship data. Journal of Applied Probability 6, 261-274.

Breslow, N. (1970). A generalized Kruskal-Wallis test for comparing K samples subject to unequal patterns of censorship. Biometrika 57, 579-594.

Breslow, N. and Crowley, J. (1974). A large sample study of the life table and product limit estimates under random censorship. Annals of Statistics 2, 437-453.

Breslow, N. and Haug, C. (1972). Sequential comparison of survival curves. Journal of the American Statistical Association 67, 691-697.

Gehan, E.A. (1965). A generalized Wilcoxon test for comparing arbitrarily single-censored samples. Biometrika 53, 203-223.

Gilbert, J.P. (1962). Random censorship. Unpublished PhD thesis, Department of Statistics, University of Chicago.

Harrington, D.P. and Fleming, T.R. (1981). A class of rank test procedures for censored survival data. Technical Report Series No. 12, Section of Medical Research Statistics, Mayo Clinic. To appear in Biometrika, 1983.

Jones, D. and Whitehead, J. (1979). Sequential forms of the logrank and modified-Wilcoxon tests for censored data. Biometrika 66, 105-113.

Kaplan, E.L. and Meier, P. (1958). Nonparametric estimation from incomplete observations. Journal of the American Statistical Association 53, 457-481.

Mantel, N. (1966). Evaluation of survival data and two new rank order statistics arising in its consideration. Cancer Chemotherapy Reports 50, 163-170.

Nagelkerke, N.J.D. and Hart, A.A.M. (1980). The sequential comparison of survival curves. Biometrika 67, 247-249.

Nelson, W. (1969). Hazard plotting for incomplete failure data. Journal of Quality Technology 1, 27-52.

Peto, R. and Peto, J. (1972). Asymptotically efficient rank invariant test procedures. Journal of the Royal Statistical Society, A 135, 185-206.

Prentice, R.L. (1978). Linear rank tests with right censored data. Biometrika 65, 167-179.

Prentice, R.L. and Marek, P. (1979). A qualitative discrepancy between censored data rank tests. Biometrics **35**, 861-867.

Slud, E.V. and Wei, L.J. (1982). Two sample repeated significance tests based on the modified-Wilcoxon statistic. Journal of the American Statistical Association. (To appear).

Tarone, R.E. (1981). On the distribution of the maximum of the logrank statistic and modified-Wilcoxon statistic. Biometrics **37**, 79-86.

Tarone, R.E. and Ware, J. (1977). On distribution free tests for equality of survival distributions. Biometrika **64**, 156-160.

Tsiatis, A.A. (1981a). A large sample study of Cox's regression model. Annals of Statistics **9**, 93-108.

Tsiatis, A.A. (1981b). The asymptotic joint distribution of the efficient scores test for the proportional hazards model calculated over time. Biometrika **68**, 311-315.

Tsiatis, A.A. (1982). Repeated significance testing for a general class of score statistics used in censored survival analysis. Journal of the American Statistical Association. (To appear).

PROCEDURES FOR SERIAL TESTING IN CENSORED SURVIVAL DATA

D.P. Harrington

Department of Applied Mathematics and Computer Science
University of Virginia, Charlottesville, Virginia

T.R. Fleming and S.J. Green

Department of Medical Statistics and Epidemiology
Mayo Clinic, Rochester, Minnesota

1. Introduction

Prospective studies, such as those carried out in many cancer centers throughout the world, need to be carefully monitored and subjected to interim analyses to satisfy important ethical considerations. Typically, the therapeutic efficacy and resulting survival distribution for an experimental treatment regimen are compared to the efficacy and survival obtained from a currently accepted standard regimen. These studies often give rise to the dual need to terminate as soon as possible any trial in which it is sufficiently clear either that (1) the experimental treatment yields better results than the standard treatment or (2) the data strongly contradict the hypothesis of some minimally acceptable treatment difference. In this paper, we examine the problem of constructing closed sequential experimental designs allowing for hypothesis tests at multiple points in time when the data gathered are censored failure time data. The tests we study are useful for examining various forms of dependence of an underlying survival function $S(x)$ on a random scalar covariate Z.

2. Model and Notation

In this manuscript we will adopt the excellent notation proposed by Tsiatis (this volume) for this problem. In particular, suppose that in a prospective study the following variables are associated with the ith study subject in a sample of n such independent and identically distributed subjects:

- Y_i: the entry time (measured from the beginning of the study)
- X_i: the time from study entry to a specified endpoint
- W_i: the time from study entry until that subject is lost to follow-up
- Z_i: random scalar valued covariate.

Since X,Y and W generally are stochastically dependent on Z (which may, for instance, denote sample membership in a two sample comparative study), the following notation will be used to denote the conditional distributions:

$$H(x|z) = P(Y \leq x|Z = z),$$
$$S(x|z) = P(X > x|Z = z),$$
$$\bar{G}(x|z) = P(W > x|Z = z).$$

We shall assume throughout that $P(Y \leq y,\ X > x,\ W > w|Z = z) = H(y|z)\ S(x|z)\ G(w|z)$ for all (y,x,w). Assume $S(x|z)$ is continuous. Let $\lambda(x|z) \equiv \frac{d}{dx}\ \ell n\ S(x|z)$.

If data of this sort are analyzed at calendar time t, that is, t units of time after the beginning of the study, the available data for subject i would include $\{X_i(t),\ \Delta_i(t),\ Z_i I\{Y_i \leq t\}\}$ where

$$X_i(t) \equiv \max\{\min(X_i,\ t - Y_i,\ W_i),\ 0\}$$

and

$$\Delta_i(t) \equiv I\{X_i \leq \min(t - Y_i,\ W_i)\}\quad .$$

Here I{E} is the indicator random variable for the event E. In many cases, these data are used to test hypotheses about the dependence of $S(x|z)$ on the covariate values z. In the next section we will review some hypotheses of interest and appropriate test statistics when one is conducting only a single test.

3. A Class of Single Stage Rank Statistics

3.1 Tests of H_0: $S(x|z) = S(x)$. The G^ρ Family

The operating characteristics of nonparametric procedures used in survival theory are most clearly understood when hypotheses are tested only once. Thus, in this section, we will temporarily assume that the data will be analyzed only at calendar time t. Without loss of generality then, we can assume in Section 3 that $H(t|z) \equiv 1$.

The most common hypothesis in this situation is $H_0: S(x|z) = S(x)$, $0 \leq x \leq t$, where $S(x)$ is unspecified. The form of the statistic used here of course depends on the way in which possible covariate dependence is modeled in the survivor function. If for the scalar covariate Z one assumes $S(x|z) = \{S_0(x)\}^{\exp(z\beta)}$, then the partial likelihood score statistic for testing the equivalent hypothesis $H_0: \beta = 0$ yields the logrank test (Mantel 1966, Cox 1972, Peto and Peto 1972). In our current notation, this statistic is

$$\sum_{i=1}^{n} \Delta_i(t) \left[Z_i - \frac{\sum_{\ell=1}^{n} Z_\ell \, I\{X_\ell(t) \geq X_i(t)\}}{\sum_{\ell=1}^{n} I\{X_\ell(t) \geq X_i(t)\}} \right].$$

A number of authors (Tarone and Ware, 1977; Prentice and Marek, 1979; and Harrington and Fleming, 1981) have proposed generalizing the above test of H_0 by incorporating weights into the terms in the above sum, yielding statistics of the form

$$(1) \qquad S_n(t) = \sum_{i=1}^{n} \hat{Q}\{t, X_i(t)\} \, \Delta_i(t) \left[Z_i - \frac{\sum_{\ell=1}^{n} Z_\ell \, I\{X_\ell(t) \geq X_i(t)\}}{\sum_{\ell=1}^{n} I\{X_\ell(t) \geq X_i(t)\}} \right].$$

The changes induced on the operating characteristics of the logrank test are clearly understood when $\hat{Q}(t,x) = \{\hat{S}(t,x)\}^{\rho}$, $\rho \geq 0$, where $\hat{S}(t,x)$ is the value at survival time $x \leq t$ of the left continuous version of the Kaplan-Meier product limit estimator computed from the pooled sample at calendar time t. This family of tests has been called the G^{ρ} family, and was proposed and studied for the k-sample problem in Harrington and Fleming (1981). Of course, when $\rho = 0$ the test is the logrank, and when $\rho = 1$, the test is essentially equivalent to the generalized Wilcoxon statistic proposed by Peto and Peto (1972) and by Prentice (1978).

For the two sample problem, where Z is 0 or 1, the following theorem indicates the types of departures against which each G^{ρ} test procedure is fully efficient. The proof relies on Corollary 5.3.1 in Gill (1980) and is given in detail in Harrington and Fleming (1981).

THEOREM 1:

Let $S_j(x) \equiv S(x|Z=j)$ and $\lambda_j(x) \equiv -\frac{d}{dx} \ln S_j(x)$ for $j = 0,1$. Fix $\rho \geq 0$. The test based on the statistic G^{ρ} is fully efficient for testing $H_0:\beta = 0$ against $H_A:\beta \neq 0$ for the Lehmann (1953) family of alternatives

(2) $$S_1(x) = S_0(x) \; [\{S_0(x)\}^{\rho} + [1 - \{S_0(x)\}^{\rho}]e^{\beta}]^{-1/\rho}, \quad 0 \leq x \leq t$$

or, equivalently,

(3) $$\lambda_1(x) = \lambda_0(x) \; e^{\beta} \; [\{S_0(x)\}^{\rho} + [1 - \{S_0(x)\}^{\rho}]e^{\beta}]^{-1}, \quad 0 \leq x \leq t$$

if and only if

$$H(t-x|Z=1) \; \bar{G}(x|Z=1) = H(t-x|Z=0) \; \bar{G}(x|Z=0), \quad x \leq t \; .$$

Interestingly, for $\rho = 0$ (resp. $\rho = 1$) we simply recover the result that the logrank test (resp. Peto and Peto - Wilcoxon test) is fully efficient for time transformed location alternatives for the extreme value (resp. logistic) distribution.

If Z is an arbitrary scalar covariate, the relationships (2) and (3) can be recast, for $\rho > 0$, as

(4) $$S(x|z) = S_0(x)\ [\{S_0(x)\}^\rho + [1 - \{S_0(x)\}^\rho]e^{\beta z}]^{-1/\rho} \quad ,$$

(5) $$\lambda(x|z) = \lambda_0(x)\ e^{\beta z}\ [\{S_0(x)\}^\rho + [1 - \{S_0(x)\}^\rho]e^{\beta z}]^{-1} \quad .$$

The G^ρ tests discussed above are applicable to testing $H_0{:}\beta = 0$ in this setting as well.

One may often be interested in testing $H_0{:}\beta = \beta_0$, β_0 not necessarily zero. This situation may arise, for instance, in cancer clinical trials when a more toxic experimental treatment is being tested against a standard treatment, and one wishes to assess whether data gathered contain significant evidence against a minimally acceptable treatment difference, say a 25% decrease in the underlying hazard. In the next two sub-sections, we will examine statistics $\tilde{S}_n(t)$ appropriate for testing the more general hypothesis $H_0{:}\beta = \beta_0$, at calendar time t. We will begin by considering the specific case of testing $H_0{:}\beta = \beta_0$ under the proportional hazards model, which warrants special consideration due to its wide applicability. This model, of course, is given in (5) when $\rho = 0$.

3.2 Tests of $H_0{:}\beta = \beta_0$ under Proportional Hazards

The proper form for $\tilde{S}_n(t)$ under the model $\lambda(x|z) = \lambda_0(x)\ e^{\beta z}$ can be seen from both a heuristic and a formal point of view. Tsiatis (1981 and this volume) has pointed out that expression (1) for $S_n(t)$, useful in testing

$H_0:\lambda(x|z) = \lambda_0(x)$, is equal to

$$S_n(t) = \sum_{i=1}^{n} \int_0^t \hat{Q}(t,x) \left[Z_i - \frac{\sum_{\ell=1}^{n} Z_\ell I\{X_\ell(t) \geq x\}}{\sum_{\ell=1}^{n} I\{X_\ell(t) \geq x\}} \right] [dN_i(t,x) - I\{X_i(t) \geq x\}\, \lambda_0(x)dx], \tag{6}$$

where $N_i(t,x) = I\{X_i(t) \leq x,\ \Delta_i(t) = 1\}$. For testing the general hypothesis $H_0:\beta = \beta_0$ under the model $\lambda(x|z) = e^{\beta z}\, \lambda_0(x)$, one might set $Q(t,x) \equiv 1$ and then replace $\lambda_0(x)$ in (6) with $e^{z_i\beta}\, \lambda_0(x)$, where, in turn, $\lambda_0(x)$ must be estimated from the data. Under $H_0:\beta = \beta_0$, and at calendar time t, a natural estimator for $\Lambda_0(x) \equiv \int_0^x \lambda_0(u)du$ is

$$\hat{\Lambda}_0(t,x) = \sum_{i=1}^{n} \int_0^x \left[\sum_{j=1}^{n} e^{\beta_0 z_j}\, I\{X_j(t) \geq u\} \right]^{-1} dN_i(t,u) \quad .$$

The following lemma, which has a simple algebraic proof, demonstrates that this heuristic approach yields a statistic which is easy to calculate.

LEMMA 1:

$$\tilde{S}_n(t) \equiv \sum_{i=1}^{n} \int_0^t \left[Z_i - \frac{\sum_{\ell=1}^{n} Z_\ell I\{X_\ell(t) \geq x\}}{\sum_{\ell=1}^{n} I\{X_\ell(t) \geq x\}} \right] \left[dN_i(t,x) - I\{X_i(t) \geq x\} e^{\beta_0 Z_i}\, d\hat{\Lambda}_0(t,x) \right] ,$$

$$= \sum_{i=1}^{n} \int_0^t \left[Z_i - \frac{\sum_{\ell=1}^{n} Z_\ell\, e^{\beta_0 Z_\ell}\, I\{X_\ell(t) \geq x\}}{\sum_{\ell=1}^{n} e^{\beta_0 Z_\ell}\, I\{X_\ell(t) \geq x\}} \right] dN_i(t,x) \quad . \tag{7}$$

Although the above approach is only intuitively reasonable, the following lemma indicates that the statistic just derived to test $H_0:\beta=\beta_0$, under the model $\lambda(x|z) = \lambda_0(x)\, e^{\beta z}$, has a more formal justification.

LEMMA 2:

Let $L(\beta,t)$ be the Cox (1975) partial likelihood constructed at calendar time t from the proportional hazards model $\lambda(x|z) = \lambda_0(x)\, e^{z\beta}$. That is,

$$L(\beta,t) = \prod_{i=1}^{n} \left[\frac{e^{Z_i\beta}}{\sum_{\ell=1}^{n} e^{Z_\ell\beta}\, I\{X_\ell(t) \geq X_i(t)\}} \right]^{\Delta_i(t)} .$$

Then $\tilde{S}_n(t)$ in expression (7) is simply the score test statistic

$$\frac{\partial}{\partial\beta} \ell n\, L(\beta,t) \Big|_{\beta=\beta_0} .$$

In this sub-section, we have discussed testing the hypothesis that

(8) $$H_0\colon \lambda(x|z) = \lambda_0(x)\, e^{\beta_0 z} .$$

It should be observed that the alternative of interest to H_0 may not always satisfy the proportional hazards assumption, i.e., may not be specified by (8) with β_0 replaced by β. More generally, if data analysis occurs at calendar time t, one may wish to test $H_0:\alpha=0$ vs $H_A:\alpha\neq 0$ where

(9) $$\lambda(x|z) = \lambda_0(x)\, \exp\{z(\beta_0+\alpha Q(t,x))\}$$

for some function $Q(t,x)$ continuous in x. We assume that $Q(t,x)$ is independent of α but may be a function of $\lambda_0(x)$. Clearly (9) reduces to (8) when $\alpha = 0$.

With analysis occurring at calendar time t, let $\hat{Q}(t,x)$ be a consistent estimator of $Q(t,x)$ under H_0. Forming Cox's partial likelihood $L(\beta_0,\alpha,t)$ based upon the relationship (9), the resulting expression can be used to formulate a score-type test statistic $\frac{\partial}{\partial\alpha} \ln L(\beta_0,\alpha,t)\Big|_{\alpha=0}$. Replacing $Q(t,x)$ by $\hat{Q}(t,x)$ yields

$$\text{(10)} \qquad \tilde{S}_n(t) = \sum_{i=1}^{n} \int_0^t \hat{Q}(t,x) \left[Z_i - \frac{\sum_{\ell=1}^{n} Z_\ell \, e^{\beta_0 Z_\ell} \, I\{X_\ell(t) \geq x\}}{\sum_{\ell=1}^{n} e^{\beta_0 Z_\ell} \, I\{X_\ell(t) \geq x\}} \right] dN_i(t,x) \; .$$

We propose $\tilde{S}_n(t)$ as defined in (10) be employed to test $H_0:\alpha = 0$ vs $H_A:\alpha \neq 0$ for the hazard relationship specified by (9). Its distribution under H_0 will be derived in §4. Observe that expression (10) reduces to (7) when $\hat{Q}(t,x) \equiv 1 \equiv Q(t,x)$ and it reduces to (1) when $\beta_0 = 0$.

Setting $\hat{Q}(t,x) = \{\hat{S}(t,x)\}^\rho$ in expression (10) would yield a generalized G^ρ statistic for testing $H_0:\alpha = 0$ in relationship (9) with

$$\text{(11)} \qquad Q(t,x) = \exp\left\{ -\rho \int_0^x \left[\frac{E\{e^{Z\beta_0} \, \bar{G}(u|Z) \; H(t-u|Z) \; \{S_0(u)\}^{\exp(Z\beta_0)}\}}{E\{\bar{G}(u|Z) \; H(t-u|Z) \; \{S_0(u)\}^{\exp(Z\beta_0)}\}} \right] \lambda_0(u)\,du \right\} .$$

When $H(u|z) = H(u)$ and $\bar{G}(u|z) = \bar{G}(u)$, equation (11) reduces to $Q(t,x) = [E\{(S_0(x))^{\exp(Z\beta_0)}\}]^\rho$. If instead one assumes $\beta_0 = 0$, then (11) reduces to $Q(t,x) = \{S_0(x)\}^\rho$. It follows that the G^ρ test procedure, specified by (10) when $\hat{Q}(t,x) = \{\hat{S}(t,x)\}^\rho$ and $\beta_0 = 0$, has been derived as being appropriate for testing $H_0:\alpha = 0$ in hazard relationship (9) with $\beta_0 = 0$ and $Q(t,x) = \{S_0(x)\}^\rho$ or, as noted earlier, for testing $H_0:\beta = 0$ in hazard relationship (5). As would be expected, these two hazard relationships are very similar.

3.3 Tests Under More General Models

We observed in the previous sub-section that the statistic in (10) arises as a score-type test statistic appropriate for testing $H_0:\alpha=0$ under the model $\lambda(x|z) = \lambda_0(x)\exp\{z(\beta_0+\alpha Q(t,x))\}$.

More generally, one could be interested in a test of the hypothesis $H_0:\alpha=0$ vs $H_A:\alpha\neq 0$ where

$$(12) \qquad \lambda(x|z) = \lambda_0(x)\exp\{z(Q_2(t,x) + \alpha Q_1(t,x))\}$$

for continuous functions $Q_i(t,x)$ which are independent of α but may be functions of $\lambda_0(x)$.

For $i=1,2$, let $\hat{Q}_i(t,x)$ be a consistent estimator under H_0 of $Q_i(t,x)$, where t continues to represent the calendar time of analysis. Forming Cox's partial likelihood based upon the relationship (12), one can again obtain a score-type statistic to test $H_0:\alpha=0$. Replacing $Q_i(t,x)$ by $\hat{Q}_i(t,x)$ yields

$$(13) \qquad \tilde{S}_n(t) = \sum_{i=1}^{n}\int_0^t \hat{Q}_1(t,x)\left[Z_i - \frac{\sum_{\ell=1}^{n} Z_\ell I\{X_\ell(t)\geq x\}\, e^{Z_\ell\hat{Q}_2(t,x)}}{\sum_{\ell=1}^{n} I\{X_\ell(t)\geq x\}\, e^{Z_\ell\hat{Q}_2(t,x)}}\right] dN_i(t,x) \quad .$$

Motivated by the frequent need, described in the Introduction, to perform interim analyses of the data, we will examine in the next section the distributions of the statistics which we have just discussed and how they can be employed when performing repeated significance testing in censored survival data. Unfortunately, it appears that the techniques to be used are only applicable when $\hat{Q}_2(t,x)$ in (13) is non-random. As a result, we will restrict our attention hereafter to statistics of the form appearing in expression (10).

4. Repeated Significance Testing in Censored Survival Data

4.1 Critical Regions for Repeated Tests

The structure of our repeated testing critical regions will be essentially that proposed by Slud and Wei (1982).

1. We will assume one will perform up to K tests based on up to n individuals. The jth test will be performed at time t_j using $\tilde{S}_n(t_j)$ as defined in (10). $\tilde{S}_n(t_j)$ is a mean zero statistic under H_0 specified by (8).

2. For a fixed overall significance level α, we will choose $\pi_1, \pi_2, \ldots, \pi_K$ such that $0 < \pi_j$ and $\sum_{j=1}^{K} \pi_j = \alpha$.

3. Critical values $\{a_1, \ldots, a_K\}$ will be recursively determined, i.e., having chosen $a_1, \ldots, a_{j-1}$, we will choose a_j so that
$$P\{\tilde{S}_n(t_1) < a_1, \ldots, \tilde{S}_n(t_{j-1}) < a_{j-1}, \tilde{S}_n(t_j) \geq a_j | H_0\} = \pi_j .$$

4. We will reject H_0 if and only if one observes the event
$$R = \bigcup_{j=1}^{K} \{\tilde{S}_n(t_1) < a_1, \ldots, \tilde{S}_n(t_{j-1}) < a_{j-1}, \tilde{S}_n(t_j) \geq a_j\},$$
which is the union of K mutually exclusive components. Thus $P(R|H_0) = \alpha$.

To carry out this approach one needs to determine the joint distribution of $n^{-1/2}\{\tilde{S}_n(t_1), \tilde{S}_n(t_2), \ldots, \tilde{S}_n(t_j)\}$ for any $t_1 \leq t_2 \leq \cdots \leq t_j$, $j = 1, \ldots, K$. We will indicate how the asymptotic joint distribution is obtained in the special case when Z_i assumes finitely many levels. Specifically, we will assume $P(Z_i = c_k) = p_k$ for $k = 1, \ldots, m$, where m is finite and $p_1 + p_2 + \cdots + p_m = 1$.

The following, which is an alternative form for $\tilde{S}_n(t)$ defined in (10) and which is the direct analogue of (6) for $\beta_0 \neq 0$, is easy to establish and will be useful in what follows.

LEMMA 3: $\tilde{S}_n(t) =$

$$\sum_{i=1}^{n} \int_0^t \hat{Q}(t,x) \left[Z_i - \frac{\sum_{\ell=1}^{n} Z_\ell e^{\beta_0 Z_\ell} I\{X_\ell(t) \geq x\}}{\sum_{\ell=1}^{n} e^{\beta_0 Z_\ell} I\{X_\ell(t) \geq x\}} \right] d\left[N_i(t,x) - \int_0^x e^{\beta_0 Z_i} I\{X_i(t) \geq u\}\, \lambda_0(u)du \right].$$

We note that $\tilde{S}_n(t)$ is of the form $\sum_{i=1}^{n} \int_0^t h_i(t,x)\, dM_i(t,x)$. For fixed t, $M_i(t,x)$ is a square integrable martingale with respect to the basis $\mathcal{F}_t^i = \{\mathcal{F}_{t,x}^i; 0 \leq x \leq t\}$, where $\mathcal{F}_{t,x}^i$ is the sigma sub-field generated by the random variables $\{I\{Y_i \leq t\}, Z_i I\{Y_i \leq t\}, Y_i I\{Y_i \leq t\}, I\{X_i \leq \min(u, t - Y_i, W_i)\}, I\{W_i \leq \min(u, t - Y_i, X_i)\}: 0 \leq u \leq x\}$. The martingale structure will be important in the following two lemmas.

LEMMA 4:

For each n and each t, let $\bar{M}_{k,n}(t,x) \equiv \sum_{i=1}^{n} M_i(t,x)\, I\{Z_i = c_k\}$; $k = 1,\ldots,m$; and let $\mathcal{B}_{n,t}$ be a basis containing $\{\mathcal{F}_t^i : i = 1,\ldots,n\}$. Let $\pi(t,x|Z) \equiv H(t-x|Z)\, S(x|Z)\, \bar{G}(x|Z)$, where we assume $\pi(t,x|Z) > 0$ for $0 < x < t$. Define

$$\hat{\mu}(t,x) = \sum_{i=1}^{n} e^{\beta_0 Z_i} Z_i I\{X_i(t) \geq x\} \Big/ \sum_{i=1}^{n} e^{\beta_0 Z_i} I\{X_i(t) \geq x\}$$

and

$$\mu(t,x) = E\left\{ e^{\beta_0 Z} Z\, \pi(t,x|Z) \right\} \Big/ E\left\{ e^{\beta_0 Z} \pi(t,x|Z) \right\},$$

where $\hat{\mu}(t,x) \equiv 0$ if $\sum_{i=1}^{n} I\{X_i(t) \geq x\} = 0$, and $\mu(t,t) \equiv 0$ if $\pi(t,t|z) = 0$. Define

$$\hat{\tilde{S}}_n(t) = \sum_{i=1}^{n} \int_0^t Q(t,x)\, \{Z_i - \mu(t,x)\}\, d\{N_i(t,x) - \int_0^x e^{\beta_0 Z_i} \lambda_0(u)\, I\{X_i(t) \geq u\} du\}.$$

Assume that

1) $\sup_{0 \le x \le \tau} |Q(t,x) - \hat{Q}(t,x)| \overset{P}{\to} 0$ for all $\tau < t$, where $\overset{P}{\to}$ means convergence in probability,

2) $\hat{Q}$ is bounded over $[0,t]$, left continuous with right hand limits, and adapted to $\mathcal{B}_{n,t}$.

Then, under H_0,

$$n^{-1/2} \{\hat{\tilde{S}}_n(t) - \tilde{S}_n(t)\} \overset{P}{\to} 0 \quad .$$

PROOF:

$$n^{-1/2} \{\tilde{S}_n(t) - \hat{\tilde{S}}_n(t)\}$$

$$\doteq \sum_{k=1}^{m} n^{-1/2} \int_0^t \{\hat{Q}(t,x) - Q(t,x)\} \{c_k - \hat{\mu}(t,x)\} \, d\bar{M}_{k,n}(t,x)$$

$$+ \sum_{k=1}^{m} n^{-1/2} \int_0^t Q(t,x) \{\mu(t,x) - \hat{\mu}(t,x)\} \, d\bar{M}_{k,n}(t,x)$$

$$\equiv \sum_{k=1}^{m} n^{-1/2} E^1_{k,n}(t) + \sum_{k=1}^{m} n^{-1/2} E^2_{k,n}(t) \quad .$$

We will establish that the above expression converges to zero in probability under H_0 by appealing to the central limit theorem (Gill 1980, §2.4) for stochastic integrals with respect to counting process martingales.

Fix ε and t. Since $\bar{M}_{k,n}$ is a square integrable martingale with respect to $\mathcal{B}_{n,t}$ (Fleming and Harrington, 1981) and $(\hat{Q} - Q)$ is bounded and predictable, $n^{-1/2} E^1_{k,n}$ is a square integrable martingale with zero expectation and predictable covariation process

$$\int \{\hat{Q}(t,x) - Q(t,x)\}^2 \{c_k - \hat{\mu}(t,x)\}^2 \, d < \bar{M}_{k,n}(t,x), \bar{M}_{k,n}(t,x) > =$$

$$\int \{\hat{Q}(t,x) - Q(t,x)\}^2 \{c_k - \hat{\mu}(t,x)\}^2 \, d \int_0^x e^{\beta_0 c_k} \lambda_0(u) \, n^{-1} \sum_{i:Z_i = c_k} I\{X_i(t) \geq u\} du \; .$$

Since sup $|\hat{Q}-Q|$ is bounded, τ_ε can be chosen such that

$$\int_{\tau_\varepsilon}^t e^{\beta_0 c_k} \lambda_0(u) du \cdot \sup_{0 \leq x \leq t} [\{Q(t,x) - \hat{Q}(t,x)\}^2 \{c_k - \hat{\mu}(t,x)\}^2] < \varepsilon/2 \; .$$

Further, since $\sup\limits_{0 \leq x \leq \tau_\varepsilon} |\hat{Q}(t,x) - Q(t,x)| \overset{P}{\to} 0$, n_1 can be chosen such that for $n \geq n_1$,

$$P(e^{\beta_0 c_k} \Lambda_0(t) \sup_{0 \leq x \leq \tau_\varepsilon} [\{\hat{Q}(t,x) - Q(t,x)\}^2 \{c_k - \hat{\mu}(t,x)\}^2] < \varepsilon/2) > 1-\varepsilon \; .$$

Then $P(< n^{-1/2} E^1_{k,n}(t), n^{-1/2} E^1_{k,n}(t) > < \varepsilon) > 1-\varepsilon$ for $n > n_1$. The martingale central limit theorem (Gill, Theorem 2.4.1) then implies that $n^{-1/2} E^1_{k,n}(t) \overset{P}{\to} 0$. We can show in a similar fashion that $n^{-1/2} E^2_{k,n}(t) \overset{P}{\to} 0$, and thus $n^{-1/2} \{\hat{\tilde{S}}_n(t) - \tilde{S}_n(t)\} \overset{P}{\to} 0$.

By Lemma 4 and an application of the Cramér-Wold device it is now sufficient to find the asymptotic joint distribtuion of $n^{-1/2}\{\hat{\tilde{S}}_n(t_1), \hat{\tilde{S}}_n(t_2), \ldots, \hat{\tilde{S}}_n(t_j)\}$. If we first define the stochastic processes $M_i(x)$, $i = 1,2,\ldots,n$

$$M_i(x) = \Delta_i(x) - \int_0^x I\{Y_i \leq u \leq Y_i + \min(X_i, W_i)\} \, \lambda(u - Y_i | Z_i) du \; ,$$

it then follows by a simple time transformation that

$$\hat{\tilde{S}}_n(t) = \sum_{i=1}^n \int_0^t Q(t, x - Y_i) \{Z_i - \mu(t, x - Y_i)\} \, dM_i(x) \equiv \sum_{i=1}^n A_i(t) \; ,$$

a sum of independent and identically distributed random variables. That $n^{-1/2}\{\hat{\tilde{S}}_n(t_1), \hat{\tilde{S}}_n(t_2),\ldots,\hat{\tilde{S}}_n(t_j)\}$ converges to a multivariate normal distribution follows from the central limit theorem. As with $M_i(t,x)$, $M_i(x)$ is a square integrable martingale, but with respect to the basis $\{\tilde{F}_x^i\ ;\ 0 \leq x \leq \infty\}$ where $\tilde{F}_x^i$ is the sigma sub-field generated by the random variables $\{I\{Y_i \leq u\}$, $Z_i I\{Y_i \leq u\}$, $I\{X_i \leq \min(u - Y_i, W_i)\}$, $I\{W_i \leq \min(u - Y_i, X_i)\}$; $0 \leq u \leq x\}$. Asymptotic moments given in the lemma below follow from results of Meyer (1976) for stochastic integrals with respect to martingales.

LEMMA 5:

Assume conditions given in Lemma 4 and let $0 \leq t \leq t' \leq t_j$. Then

$$E\,\hat{\tilde{S}}_n(t) = E\,A_i(t) = 0, \text{ and } \operatorname{cov}\{n^{-1/2}\,\hat{\tilde{S}}_n(t),\ n^{-1/2}\,\hat{\tilde{S}}_n(t')\} =$$

$$\operatorname{cov}\{A_i(t),A_i(t')\} = \int_0^t Q(t,x)\ Q(t',x)\,E[\{Z-\mu(t,x)\}^2\ e^{\beta_0 Z}\ \lambda_0(x)\ \pi(t,x|Z)]\,dx\ .$$

When $\beta_0 = 0$, one obtains results presented by Tsiatis (this volume). However, application of martingale stochastic integral results simplifies the covariance calculation he made for this special case.

Since $\operatorname{cov}\{A_i(t),A_i(t')\}$ depends upon t' only through $Q(t',x)$, it follows that $\{n^{-1/2}\ \tilde{S}_n(t){:}\,t \geq 0\}$ converges to a limit process having independent increments whenever $Q(t,x)$ is independent of t. Such is the case for the generalized G^ρ family when either $S(x|Z)$ or $H(x|Z)$ is independent of Z. This can be seen by observing $Q(t,x) = \{S^*(t,x)\}^\rho$ when $\hat{Q}(t,x) = \{\hat{S}(t,x)\}^\rho$, where

$$S^*(t,x) \equiv \exp[-\int_0^x \left\{ \sum_{k=1}^m p_k\ \pi(t,u|Z{=}c_k)\,\lambda(u|Z{=}c_k) \Big/ \sum_{k=1}^m p_k\ \pi(t,u|Z{=}c_k) \right\} du]\ .$$

Although the derivation of the asymptotic distribution of $n^{-1/2}\{\tilde{S}_n(t_1),\ldots, \tilde{S}_n(t_k)\}$ assumes independent, identically distributed covariates Z_i, it can be extended to studies in which the covariate values are balanced through forced randomization. In particular, assume that in a study of n subjects, n_k of those will have covariate value c_k, $k=1,\ldots,m$ and that $n_k \equiv np_k$. The term $\sum_{i=1}^{n} A_i(t)$ may then be viewed as a sum of independent, but non-identically distributed terms. One may still show that $n^{-1/2}\{\tilde{S}_n(t_1),\ldots,\tilde{S}_n(t_k)\}$ is asymptotically multivariate normal, with zero mean and with

$$\lim_{n\to\infty} \mathrm{Cov}\{n^{-1/2}\,\tilde{S}(t),\, n^{-1/2}\,\tilde{S}_n(t')\}$$

$$= \sum_{k=1}^{m} p_k \int_0^t Q(t,x)\; Q(t',x)\; \{c_k - \mu(t,x)\}^2\; e^{\beta_0 c_k}\; \lambda_0(x)\; \pi(t,x|c_k)dx\ ,$$

where $t<t'$. Note that this expression agrees with the covariance formula in Lemma 5.

Recall, by Lemma 5, that $\mathrm{cov}\{n^{-1/2}\,\tilde{S}_n(t),\, n^{-1/2}\,\tilde{S}_n(t')\}$ converges to

$$\sigma(t,t') \equiv \int_0^t Q(t,x)\; Q(t',x)\; E[\{Z-\mu(t,x)\}^2\, e^{\beta_0 Z}\; \pi(t,x|Z)]d\Lambda_0(x)\quad .$$

If $\hat{\sigma}(t,t')$ denotes a consistent estimator of $\sigma(t,t')$, then the actual test statistics employed at t and t' are $n^{-1/2}\{\hat{\sigma}(t,t)\}^{-1/2}\,\tilde{S}_n(t)$ and $n^{-1/2}\{\hat{\sigma}(t',t')\}^{-1/2}\,\tilde{S}_n(t')$, which are positively correlated. One consistent estimator is given by $\hat{\sigma}(t,t') \equiv$

$$\int_0^t \hat{Q}(t,x)\; \hat{Q}(t',x)\left[\frac{1}{n}\sum_{j=1}^{n} \{Z_j-\hat{\mu}(t,x)\}^2\; e^{\beta_0 Z_j}\; I\{X_j(t)\geq x\}\right] d\hat{\Lambda}_0(t,x) =$$

$$\frac{1}{n}\sum_{i\varepsilon R(t)}\Bigg[\Delta_i(t)\;\hat{Q}(t,X_i(t))\;\hat{Q}(t',X_i(t))$$

$$*\left\{\sum_{j\varepsilon R(t,X_i(t))} e^{\beta_0 Z_j}\left\{Z_j - \frac{\sum_{\ell\varepsilon R(t,X_i(t))} Z_\ell\, e^{\beta_0 Z_\ell}}{\sum_{\ell\varepsilon R(t,X_i(t))} e^{\beta_0 Z_\ell}}\right\}^2 \Bigg/ \sum_{j\varepsilon R(t,X_i(t))} e^{\beta_0 Z_j}\right\}\Bigg],$$

where R(t) denotes the set of indices $\{j=1,\ldots,n\}$ such that $Y_j \leq t$, and R(t,x) denotes the set such that $\{X_j(t) \geq x\}$. That $n^{-1/2}\{\hat{\sigma}(t,t)\}^{-1/2}\, \tilde{S}_n(t)$ does not require knowledge of n is important for applications.

4.2 Selection of π_i, $i=1,\ldots,K$.

Several different approaches exist for choosing π_i; $i=1,\ldots,K$. One approach of particular interest would be to select $\pi_1, \pi_2,\ldots,\pi_{K-1}$ small, with $\pi_K \approx \alpha$, where α is the size of the procedure. Procedures discussed by Haybittle (1971) and O'Brien and Fleming (1979) are conceptually related to this. The resulting serial testing procedure would then allow early testing to detect substantial departures from H_0, satisfying ethical considerations. In addition, the critical value for the statistic employed at the Kth and final stage of the procedure would be nearly identical to the critical value which is appropriate when a single procedure is based upon that statistic. Such a sequential procedure would have power nearly identical to that of the corresponding single stage procedure. On the other hand, the serial testing procedure resulting from repeated use of a statistic will have operating characteristics considerably different from those of the corresponding single stage procedure if one chooses π_i, $i=1,\ldots,K$, such that $\pi_K \ll \alpha$. In heavily censored data, the sequential procedure will give much ligher weight to "later" occurring departures from H_0 than the corresponding single stage procedure. A careful theoretical consideration of the power and efficiencies of these types of serial testing procedures seems to be difficult.

ACKNOWLEDGEMENT

The research of the first author was supported by Eagles Grant 23 and NSF Grant MSC-80-02887. The research of the second and third authors was supported by NIH Grant CA-24089.

REFERENCES

Cox, D.R. (1972). Regression models and life tables (with discussion). Journal of the Royal Statistical Society B **34**, 187-220.

Cox, D.R. (1975). Partial likelihood. Biometrika **62**, 269-276.

Fleming, T.R. and Harrington, D.P. (1981). A class of hypothesis tests for one and two sample censored survival data. Communications in Statistics A **10**, 763-794.

Gill, R.D. (1980). Censoring and Stochastic Integrals. Mathematical Centre Tracts 124, Mathematische Centre, Amsterdam.

Haybittle, J.L. (1971). Repeated assessment of results in clinical trials of cancer treatment. British Journal of Radiology **44**, 793-797.

Harrington, D.P. and Fleming, T.R. (1981). A class of rank test procedures for censored survival data. Technical Report Series No-12, Section of Medical Research Statistics, Mayo Clinic. To appear in Biometrika, 1983.

Lehmann, E.L. (1953). The power of rank tests. Annals of Mathematical Statistics **24**, 23-43.

Mantel, N. (1966). Evaluation of survival data and two new rank order statistics arising in its consideration. Cancer Chemotherapy Reports **50**, 163-170.

Meyer, P.A. (1976). Un cours sur les integrales stochastiques, p. 245-400. In: Seminaire de Probabilities X, Lecture Notes in Mathematics **511**, Springer-Verlag, Berlin.

O'Brien, P.C. and Fleming, T.R. (1979). A multiple testing procedure for clinical trials. Biometrics **35**, 549-556.

Peto, R. and Peto, J. (1972). Asymptotically efficient rank invariant test procedures (with discussion). Journal of the Royal Statistical Society A **135**, 185-206.

Prentice, R.L. (1978). Linear rank tests with right censored data. Biometrika **65**, 167-179.

Prentice, R.L. and Marek, P. (1979). A qualitative discrepancy between censored data rank tests. Biometrics **35**, 861-867.

Slud, E.V. and Wei, L.J. (1982). Two-sample repeated significance tests based on the modified Wilcoxon statistic. Journal of the American Statistical Association. To appear.

Tarone, R.E. and Ware, J. (1977). On distribution free tests for equality of survival distributions. Biometrika **64**, 156-160.

Tsiatis, A.A. (1981a). The asymptotic joint distribution of the efficient scores test for the proportional hazards model calculated over time. Biometrika **68**, 311-315.

SIMULATION STUDIES ON INCREMENTS OF THE TWO-SAMPLE LOGRANK SCORE TEST FOR SURVIVAL TIME DATA, WITH APPLICATION TO GROUP SEQUENTIAL BOUNDARIES

Mitchell H. Gail

Biometry Branch, National Cancer Institute, Bethesda, Maryland

David L. DeMets

Mathematical and Applied Statistics Branch, National Heart, Lung and Blood Institute, Bethesda, Maryland

Eric V. Slud

Department of Mathematics, University of Maryland, College Park, Md.

0. SUMMARY

The performance of the logrank statistic, computed after successive fixed numbers of deaths and applied to group sequential boundaries, is evaluated using simulation studies. The group sequential boundaries investigated include those proposed by Haybittle (1971), Pocock (1977), O'Brien and Fleming (1979) and the fixed sample boundary. The data indicate that a simple normal model, based on the assumptions that the increments of the logrank score are uncorrelated and homoscedastic with known variance, leads to reliable predictions of size, power, and average number of groups examined, except when the numbers at risk are very small, as in completely sequential entry. When there is a trend in the lifetime distribution, either in location or dispersion, the size of some group sequential boundaries exceeds nominal levels slightly, whereas the fixed sample logrank test is robust to such trends. The assumptions that the logrank increments are uncorrelated and homoscedastic with known variance are also investigated.

1. Introduction

Pocock (1977) proposed to monitor accumulating clinical trial data with group sequential boundaries which are appropriate for repeated analyses after successive groups of observations. After n groups of d observations each, the standardized statistic $T(nd) = (\sum_{i=1}^{n} \sum_{j=1}^{d} Y_{ij})(\sigma^2 nd)^{-\frac{1}{2}}$ is computed and compared with symmetric, two-sided group sequential boundaries c_n for $n=1,2,\ldots,N$. The null hypothesis is rejected at the smallest $n < N$ for which $|T(nd)| > c_n$. To compute boundaries of appropriate size, it is assumed that the group increments $\sum_{j=1}^{d} Y_{ij}$ are normally distributed, uncorrelated, and homoscedastic with known variance $\sigma^2 d$. Power is computed under the alternative that the group increments have mean δd.

Recently Pocock (1980) suggested that such boundaries could be applied to comparative survival studies by analyzing the standardized logrank statistic at intervals defined by equal numbers of deaths. This idea is closely related to suggestions by Armitage (1975, p. 143) and Jones and Whitehead (1979) for fully sequential analyses. Rigorous asymptotic theory in support of this proposal is available for two special cases, namely, progressive censorship, in which all patients enter simultaneously at the beginning of the experiment, and completely sequential entry, in which the lifetime of one patient is determined before the next enters the study. By referring to the permutational distribution of the linear rank statistic, Chatterjee and Sen (1973) showed that increments of the logrank score (numerator) are uncorrelated under progressive censorship, even for small samples. Asymptotically, their results imply that these increments are normally distributed and homoscedastic with known variance. Sen and Ghosh (1972) obtained these results for sequential entry.

The purpose of our simulations was to cover the intermediate case of staggered entry, which is of practical concern. Even for progressive censorship and sequential entry, simulations were useful to indicate the extent to which asymptotic theory applied. For staggered entry, Tsiatis (1981) has shown the increments to be asymptotically uncorrelated when the intervals are defined by fixed calendar times rather than by fixed numbers of deaths.

We report on the operating characteristics of group sequential boundaries proposed by Haybittle (1971), Pocock (1977) and O'Brien and Fleming (1979). In addition, we provide data on the correlation structure of the increments of the logrank score and test the hypotheses H1 that the increments are uncorrelated, H2 that the increments are uncorrelated and homoscedastic, and H3 that the increments are uncorrelated and homoscedastic with known variance d/4.

Some special studies were undertaken to determine whether group sequential procedures are robust to trends in the life distribution.

2. Methods

2.1 Definition of the Statistics and Boundaries

The computation of the two-sample logrank statistic is particularly simple in the case of continuous survival data (no ties) which we treat. As in Mantel (1966) and Cox (1972), we order the death times $t_1 < t_2 < \cdots < t_d$ to compute the logrank statistic after d deaths. Let p_k denote the proportion of all those patients known to have survived for time t_k or longer who are in group 1, and let $U_k = 1$ or 0 according as the death at t_k is in group 1 or 2. Then the logrank score after d deaths is

$$Z(d) = \sum_{k=1}^{d} (U_k - p_k) .$$

The estimated variance of Z(d) is

$$V(d) = \sum_{k=1}^{d} p_k(1 - p_k) ,$$

which is only slightly less than d/4 in most cases with equal allocation, provided treatment effects are not too large.

After n groups (nd deaths), the statistic

$$T(nd) = Z(nd) \{V(nd)\}^{-\frac{1}{2}}$$

is computed and compared to symmetric two-sided group sequential boundaries c_n for $n = 1,2,\ldots,N$, where N is the maximum number of groups to be entered. Tests proceed in the manner of Pocock (1977), with a rejection decision reached for the smallest n such that

$$|T(nd)| > c_n, \quad n = 1,2,\ldots,N \quad .$$

We examined four symmetric two-sided size $\alpha = 0.05$ boundaries and studied the case $N = 5$ in detail. For $N = 5$, the Pocock (1977) boundary (P) is $c_n = 2.413$ for $n = 1,2,\ldots,5$. The Haybittle (1971) boundary (H) is $c_n = 3.0$ for $n = 1,2,3,4$ and $c_5 = 1.96$. This boundary is conservative and only detects extreme early treatment differences. The O'Brien-Fleming (1979) boundary (O), obtained from their Table 1, is $c_n = (4.149 \times 5/n)^{\frac{1}{2}}$ for $n = 1,2,\ldots,5$. For the fixed sample boundary (F), $c_n = 100$ for $n = 1,2,3,4$ and $c_5 = 1.96$. The value $c_n = 100$ was chosen for convenience and was never exceeded in our studies.

Suppose Z(nd) were the cumulative sum of nd independent Bernoulli variates corresponding to factors in the partial likelihood of Cox (1972), and that these factors were essentially unaffected by survival information obtained after death nd. Then under the null hypothesis of equality of survival distributions, increments such as $Z(2d) - Z(d)$ would have expectation zero, and correlation zero. With equal allocation, the variance of these increments is approximately d/4. For proportional hazards alternatives with hazard ratio $\exp(\theta)$, the expectation of such an increment would be approximately $d\theta/4$ for small θ. Thus, the power can be expressed in terms of the group non-centrality parameter

$$(1) \qquad \Delta = (d\theta/4)(d/4)^{-\frac{1}{2}} = \theta(d/4)^{\frac{1}{2}} \quad .$$

This quantity Δ is used for tabulations in Pocock (1977). Assuming the normal model holds, the theoretical size, power, and average number of groups $\bar{n}$ may be computed as in Armitage, McPherson and Rowe (1969), McPherson and Armitage (1971), and DeMets and Ware (1980). The theoretical variance of the stopping number n may also be determined from the multinomial distribution of the stopping points.

2.2 Description of the Simulation

The simulated clinical trials had a maximum of Nd = 90 deaths. This number of deaths was determined from equation (1) with N = 1 group so as to yield a power 0.90 for the two-sided 0.05 level logrank test with boundary F against the alternative of a two-fold relative hazard. Lininger, et al, (1979) confirm by simulations that equation (1) indeed yields the correct numbers of deaths required to attain specified power with the boundary F. The simulation proceeded by generating Poisson entry times for each patient up to a maximum of M = 90, 135 or 180 patients. The case M = 90 requires all patients to be followed to death. The case M = 180 allows the trial to stop when 90 patients are either still at risk or yet to be accrued, depending on the rate of entry. Larger values of M were not considered because, in the presence of rapid accrual, the decision would be reached on the basis of early deaths only, and in the case of slow accrual, increasing M beyond 180 has little effect. Each entered patient was then assigned a treatment using an independent Bernoulli variate (usually with equal allocation) and a lifetime (usually exponential). Exponential lifetimes and Poisson entry waiting intervals were generated with the IMSL subroutine GGEXN, and Bernoulli variates were based on the uniform IMSL pseudorandom numbers from GGUBFS. The IBM 370/OSVS was used. At the time when nd deaths occurred, the logrank statistic was calculated by resorting the follow-up times of all patients who had entered the trial to that time. The case of progressive censoring was studied by letting the Poisson entry rate get very large, and the case of sequential entry was studied by letting the entry rate get very small. Each experimental design was studied using 1000 independent simulations of the clinical trial. The proportion of rejections for the empirical estimates of size and power in Tables 2 and 3 are shown as the number of rejections per 1000 trials. Results for each boundary studied are correlated within each experimental design, but all statisitcs were independent across designs.

For Poisson entry experiments with $N = 5$, the uncorrelated increments assumption (H1) is tested using the normal theory likelihood ratio statistic computed from formula 7, page 239 in Anderson (1958). The test of H2, the assumptions of uncorrelated homoscedastic increments, is given by equation 7 on page 261 in Anderson. The assumption H3 of independent increments with common variance $d/4$ is tested using equation 7, page 265 in Anderson.

3. Results

3.1 Theoretical Properties of the Group Sequential Boundaries Based on the Normal Model

The results of Table 1 were computed under the normal model using the numerical methods described by DeMets and Ware (1980). The noncentrality parameter Δ was computed from (1) with $d = 18$, $N = 5$, and relative hazard $\exp(\theta) = 2$. The power of H exceeds that of F only because H has size 0.053, slightly in excess of 0.05. The Pocock boundary offers the greatest average savings in n but also has the least power.

TABLE 1. Theoretical properties of four group sequential boundaries with $N = 5$

	Haybittle (H)	Pocock (P)	Fixed (F)	O'Brien-Fleming (O)
Null case $\Delta = 0$				
size	0.053	0.050	0.050	0.050
$\bar{n}$	4.977	4.876	5.000	4.964
SD(n)	0.268	0.622	0.000	0.241
Alternative $\Delta = 1.470$				
power	0.909	0.845	0.907	0.901
$\bar{n}$	3.864	3.083	5.000	3.648
SD(n)	1.313	1.441	0.000	0.989

3.2 Null Case Results with $N = 5$

The null case data in Table 2 are generated using unit exponential lifetimes in both treatment groups. Most, but not all, experimental conditions in Table 2 were studied in two independent simulations. These data give no evidence against the uncorrelated increments assumption H1. Evidence against homoscedasticity (H2) and/or known common variance equal to $d/4$ (H3) is seen when patients enter sequentially (entry rate 0.001 per year) and when only $M = 90$ patients are admitted. Both these cases require that observation continue until the last patient dies. These are, of course, situations in which p_k may deviate from 0.5 and in which incremental variances $V(nd) - V\{(n-1)d\}$ may deviate markedly from $d/4$. The smaller variance which results when p_k deviates from 0.5 may account for the slight but consistent elevations in size above nominal levels observed for sequential entry. This holds for all boundaries but is especially pronounced for the Pocock boundary P which has average size 0.069 for sequential entry. The deviation $\delta\bar{n}$ of the average number of groups $\bar{n}$ from predicted is shown for the Pocock boundary. The observed $\bar{n}$ for the Pocock boundary is in good agreement with the predicted value 4.876 except for the case of sequential entry where $\bar{n}$ is slightly smaller than predicted. To summarize, these null experiments are consistent with the uncorrelated increments assumption, and, except for cases when p_k may deviate markedly from 0.5, the homoscedasticity and known variance assumptions are also tenable. The size and average sample number of these experiments are consistent with theory based on the normal model except for minor discrepancies in the case of sequential entry.

3.3 Non-Null Results with $N = 5$

The siutation is different under the alternative hypothesis with exponential lifetime hazards 2.0 and 1.0 in the two treatment groups. The nested hypotheses H1, H2 and H3 are often violated.

The non-null operating characteristics of the group sequential boundaries are detectably different from predictions of the normal model, but the effects are not gross and not of practical importance. The power of the Pocock

TABLE 2. Simulations with N = 5 based on 1000 repetitions for each experiment

NULL CASE

Total Patients M	Progressive Censoring 100,000/Year						Fast Staggered Entry 100/Year					
	H1	H2	H3	$H^{\dagger}$ 0	P F	Pocock $\delta\overline{n}$**	H1	H2	H3	H 0	P F	Pocock $\delta\overline{n}$
				59	53					58	53	
90	12	36*	38*	52	56	-.003	7	13	13	59	54	.002
				55	52					59	53	
90	16	39*	46*	46	49	-.023	7	30*	38*	57	58	.008
				47	49					53	49	
135	10	14	14	50	45	.011	15	18	21	46	49	.006
				47	45					37	42	
135	11	16	19	45	44	.006	9	15	21	37	33	.005
				40	48					51	55	
180	10	14	17	39	38	-.008	6	10	17	49	47	-.028
				45	44					54	43	
180	6	6	8	40	43	.016	8	12	14	51	52	.012

NULL CASE

Total Patients M	Slow Staggered Entry 10/Year						Sequential Entry 0.001/Year					
	H1	H2	H3	H 0	P F	Pocock $\delta\overline{n}$	H1	H2	H3	H 0	P F	Pocock $\delta\overline{n}$
				51	57					67	78	
90	4	13	23	51	47	-.022	14	20	24	59	59	-.089
				57	63					73	74	
90	9	19	32*	57	52	-.039	10	29*	31*	54	63	-.094
				49	48					63	67	
135	7	22	32*	48	42	-.011	11	27*	32*	55	56	-.045
				49	51					67	64	
135	2	16	31*	51	46	-.013	7	36*	44*	66	56	-.050
				46	60					67	69	
180	3	9	19	39	42	-.034	4	27*	35*	62	62	-.063
				53	52					67	63	
180	6	14	37*	56	47	-.021	7	18	26*	57	56	-.059

Table 2 (continued)

RELATIVE HAZARD 2

Progressive Censoring 100,000/Year | Fast Staggered Entry 100/Year

Total Patients	H1	H2	H3	H_0	P_F	Pocock $\delta\bar{n}$	H1	H2	H3	H_0	P_F	Pocock $\delta\bar{n}$
				889	811					889	813	
90	152*	441*	459*	875	889	.066	168*	382*	413*	891	889	.038
				877	794					884	811	
90	179*	531*	542*	863	875	.087	73*	298*	316*	869	882	.058
				908	841					896	842	
135	17	24*	29*	896	908	.010	9	15	17	890	895	.088
				909	851					905	837	
135	13	25*	28*	895	906	.012	17	24*	26*	901	904	.056
				900	828					898	819	
180	16	22	32*	891	900	.162	3	10	11	883	897	.102
180												

RELATIVE HAZARD 2

Slow Staggered Entry 10/Year | Sequential Entry 0.001/Year

Total Patients	H1	H2	H3	H_0	P_F	Pocock $\delta\bar{n}$	H1	H2	H3	H_0	P_F	Pocock $\delta\bar{n}$
				889	811					892	800	
90	38*	63*	162*	883	886	.177	12	22	284*	876	890	.234
				884	812					892	825	
90	35*	52*	144*	863	880	.082	27*	37*	264*	887	891	.163
				915	841					891	818	
135	49*	50*	123*	905	913	.016	13	16	240*	883	888	.209
				899	830					882	822	
135	34*	38*	80*	877	888	.165	13	15	247*	875	881	.114
				900	816					892	835	
180	40*	44*	93*	882	896	.099	12	21	321*	882	889	.075
				899	805							
180	53*	57*	168*	892	899	.207						

* Exceeds the 95th percentile of the corresponding chi-square distribution. The degrees of freedom are 10 for H1, 14 for H2 and 15 for H3.

† Size and power estimates are given for each boundary as the number of rejections per 1000 repetitions.

**The quantity $\delta\bar{n}$ is the deviation of the average number of increments, $\bar{n}$, from expectation. For the null case, $\delta\bar{n} = \bar{n} - 4.876$, and for the relative hazard 2, $\delta\bar{n} = \bar{n} - 3.083$.

boundary is less than the theoretical value 0.845 in every experiment but one and tends to decrease as the entry rate decreases. Nonetheless, the observed power is usually only about 3% less than predicted.

Under the two-fold hazard ratio alternative, the observed values of $\bar{n}$ for the Pocock boundary tend to exceed the predicted value 3.083, especially for entry rate 0.001 per year. These discrepancies range from 0.3 to 7.6% of predicted and are tolerable in practice. For the other boundaries, which have less potential for early stopping, the discrepancies are even smaller.

3.4 Miscellaneous Experiments

A few staggered entry experiments were conducted with $N = 10$, $d = 9$. These experiments conform to H1, H2 and H3 and to the normal model predictions even better than results in Table 2. Some experiments were performed with Weibull lifetimes. For shape parameter 3, null and non-null results were similar to those in Table 2 (entry rate 10/year). With shape parameter 1/3, null and non-null conformance to the normal model was better than indicated in Table 2.

3.5 Robustness Studies When There are Trends in the Life Distribution

The first eight experiments in Table 3 reflect the performance of group sequential boundaries with $N = 5$ and $M = 135$ when there is a time trend in the mean exponential life. The mean lifetime varies linearly with the patient entry index $\gamma = i/135$ for $i = 1,2,\ldots,135$. To obtain a simulated lifetime in the first experiment, a simulated unit exponential lifetime ℓ_0 is transformed to $\ell = (0.1 + 0.9\gamma)\ell_0$. Thus, the average exponential lifetime increases about 10 fold as i ranges from 1 to 135. Other trends are produced from $\ell = (0.5 + 0.5\gamma)\ell_0$, $\ell = (1.0 - 0.5\gamma)\ell_0$, and $\ell = (1.0 - 0.9\gamma)\ell_0$. Under the null hypothesis of no treatment effect, the lifetimes being compared are from the same mixed exponential population. With sequential entry, these null case experiments demonstrate that the increments are negatively correlated for increasing mean life trends and positively correlated for strongly decreasing mean life trends.

TABLE 3. Null case robustness studies with trends in the life distribution for N = 5, M = 135

Description of the trend	H1	H2	H3	Size P	Size O	Size H	Size F
Linear trend in the mean exponential life							
Sequential entry (0.001/yr)							
10 fold increase	143*	370*	896*	82	68	69	55
2 fold increase	51*	135*	192*	71	48	53	40
2 fold decrease	12	33*	112*	67	55	58	51
10 fold decrease	42*	86*	317*	78	57	65	56
Staggered entry (10/yr)							
10 fold increase	122*	291*	690*	86	64	69	56
10 fold decrease	29*	56*	173*	52	48	55	50
Progressive censorship (10^5/yr)							
10 fold increase	6	9	9	50	54	51	48
10 fold decrease	5	9	9	38	47	52	52
Trend in dispersion with constant geometric mean							
Sequential entry (0.001/yr)							
Base 10 increase	69*	219*	363*	69	50	56	45
Base 2 increase	15	42*	44*	65	47	58	49
Base 1	6	16	26*	67	52	63	55
Base 2 decrease	15	20	82*	55	44	50	42
Base 10 decrease	100*	215*	554*	63	44	56	50
Progressive censorship (10^5/yr)							
Base 10 increase	5	15	17	56	63	62	66
Base 10 decrease	10	12	12	51	65	72	70
Trend in mean and dispersion of uniform lifetimes							
Sequential entry (0.001/yr)							
U (12, 12+10γ)	558*	1033*	3062*	99	77	64	49
U (12, 12-10γ)	258*	353*	990*	62	52	56	51

*Exceeds the 95th percentile of the corresponding chi-square distribution.

The size of the Pocock boundary exceeds 0.05 in these cases. These effects are still appreciable at staggered entry rate 10 per year, but they vanish for rapid entry, which is to be anticpated from the theory of progressive censorship applied to a mixed exponential population. Note that the fixed sample test F, and the conservative boundaries O and H, are more robust to such trends.

Seven experiments investigate the effects of a trend in dispersion with constant geometric mean. For the first of these, a unit exponential lifetime ℓ_0 is changed to $\ell = \ell_0 \times 10^{\gamma}$ with probability 1/3, to $\ell = \ell_0 \times 10^{-\gamma}$ with probability 1/3 and to $\ell = \ell_0$ otherwise. In this transformation, 10 is the "base" and $\gamma = i/135$ as before. This yields increasing dispersion. For decreasing dispersion, γ is replaced by $1 - \gamma$. The effects on size and correlation structure are smaller than for a trend in the exponential mean.

Rather dramatic effects on size and correlation structure are seen for trends in the lifetimes which are assumed to be uniform on the support interval indicated. The size of the Pocock boundary is 0.099 in one instance. Again, note the robustness of the fixed sample procedure F.

3.6 Some Null Case Experiments with Two Batches of Patients

We shall briefly mention the results of a number of null case experiments in which a first batch of 100 patients entered at time zero and a second batch of 100 patients entered at the time of death $d = 20$ in the first batch. Group sequential analyses were performed at deaths $d = 20$ and $d = 40$. In some experiments, batches consisted of 50 patients. The principal conclusions of these experiments were:

(1) Loss to follow-up, as might occur if a patient refuses further participation, does not affect size. Loss to follow-up was studied by assuming a constant hazard of withdrawal from the study after the patient is entered. It was found that size is not affected by unequal loss to follow-up in the two treatment groups. Again, size remains near 0.05 even if the loss to follow-up is adaptive in the sense that

the risk of loss to follow-up in the second batch is greater on the treatment which appeared worse at the first analysis. This latter situation could arise in practice if patients who enter learn which treatment appears to be more successful and adhere preferentially to the favored treatment.

(2) Adaptive allocation of 80% of the second batch to that treatment which appreared better (or worse) at the time of the first analysis does not affect size.

(3) Whether the second batch has longer or shorter mean lifetimes than the first batch, size remains near 0.05. If the second batch has a mean lifetime 10 fold greater than the first batch, the logrank score increments are negatively correlated. Interestingly, even if the second batch has a mean lifetime 1000 fold smaller than the first batch, the increments appear uncorrelated. Thus, distortions are more prominent when healthy patients enter later. This asymmetry is also seen in the first six experiments of Table 3. There too, a trend toward healthier patients induces stronger correlations among increments of the logrank score than does a trend toward sicker patients.

4. DISCUSSION

The correlation structure of the logrank score statistic conforms rather well to the normal model (H3) under the null hypothesis except for purely sequential entry. Departures from this simple correlation structure are evident under the alternative of a two-fold hazard ratio and when there are trends in the life distribution. It is not surprising, therefore, that theoretical calculations of operating characteristics work best under the null hypothesis. What is noteworthy is that these calculations are sufficiently accurate to plan experiments under the alternative. For any size α boundary $\{c_n\}$, one can calculate the non-centrality parameter required to obtain a desired power as in DeMets and Ware (1980) or Pocock (1977). Then the required number of deaths per increment is obtained from (1).

Robustness studies demonstrate that the size of the Pocock boundary slightly exceeds 0.05 when there are trends in the lifetime distribution. A typical empirical size is about 0.07. As expected, the boundaries O and H are less affected by such trends, and F is completely robust.

One can adhere to a group sequential plan by analyzing the data when prespecified numbers of deaths have occurred. If one plans to perform repeated analyses at fixed calendar times, instead, predetermined boundaries may turn out to be inappropriate because accrual and death rates are variable and hard to predict.

ACKNOWLEDGEMENTS

We wish to thank James H. Ware for helpful suggestions and Julie Paolella for typing this manuscript.

REFERENCES

Anderson, T.W. (1958). An introduction to multivariate statistical analysis. New York: Wiley.

Armitage, P. (1975). Sequential medical trials. New York: Wiley.

Armitage P., McPherson, C.K., and Rowe, B.C. (1969). Repeated significance tests on accumulating data. Journal of the Royal Statistical Society A **132**, 235-244.

Chatterjee, S.K., and Sen, P.K. (1973). Non-parametric testing under progressive censoring. Calcutta Statistical Association Bulletin **22**, 13-50.

Cox, D.R. (1972). Regression models and life tables (with discussion). Journal of the Royal Statistical Society B **34**, 187-220.

DeMets, D.L. and Ware, J.H. (1980). Group sequential methods in clinical trials with a one-sided hypothesis. Biometrika **67**, 651-660.

Habittle, J.L. (1971). Repeated assessment of results in clinical trials of cancer treatment. British Journal of Radiology **44**, 793-797.

Jones, D., and Whitehead, J. (1979). Sequential forms of the logrank and modified Wilcoxon tests for censored data. Biometrika 66, 105-113.

Lininger, L., Gail, M.H., Green, S.B., and Byar, D.P. (1979). Comparison of four tests for equality of survival curves in the presence of stratification and censoring. Biometrika 66, 419-428.

Mantel, N. (1966). Evaluation of survival data and two new rank order statistics arising in its consideration. Cancer Chemotherapy Reports 50, 163-170.

McPherson, C.K., and Armitage, P. (1971). Repeated significance tests on accumulating data when the null hypothesis is not true. Journal of the Royal Statistical Society A 134, 15-25.

O'Brien, P.C. and Fleming, T.R. (1979). A multiple testing procedure for clinical trials. Biometrics 35, 549-556.

Pocock, S.J. (1977). Group sequential methods in the design and analysis of clinical trials. Biometrika 64, 191-199.

Pocock, S.J. (1980). Group sequential design for clinical trials. Presented at the American Statistical Association Meetings in Houston, August 11-14.

Sen, P.K. and Ghosh, M. (1972). On strong convergence of regression rank statistics, Sankhya A 34, 335-348.

Tsiatis, A.A. (1981). The asymptotic joint distribution of the efficient scores test for the proportional hazards model calculated over time. Biometrika 68, 311-315.